Lectures on
Fractal Geometry

Fractals and Dynamics in Mathematics, Science, and the Arts: Theory and Applications

ISSN: 2382-6320

Published

For the complete list of volumes in this series, please visit
www.worldscientific.com/series/fds

Lectures on Fractal Geometry

Martina Zähle

University of Jena, Germany

World Scientific

NEW JERSEY • LONDON • SINGAPORE • BEIJING • SHANGHAI • HONG KONG • TAIPEI • CHENNAI • TOKYO

Published by

World Scientific Publishing Co. Pte. Ltd.

5 Toh Tuck Link, Singapore 596224

USA office: 27 Warren Street, Suite 401-402, Hackensack, NJ 07601

UK office: 57 Shelton Street, Covent Garden, London WC2H 9HE

Library of Congress Control Number: 2023047450

British Library Cataloguing-in-Publication Data
A catalogue record for this book is available from the British Library.

Fractals and Dynamics in Mathematics, Science, and the Arts:
Theory and Applications — Vol. 8
LECTURES ON FRACTAL GEOMETRY

ISBN 978-981-128-333-8 (hardcover)
ISBN 978-981-128-334-5 (ebook for institutions)
ISBN 978-981-128-335-2 (ebook for individuals)

For any available supplementary material, please visit
https://www.worldscientific.com/worldscibooks/10.1142/13594#t=suppl

Desk Editors: Aanand Jayaraman/Ana Ovey

Typeset by Stallion Press
Email: enquiries@stallionpress.com

Printed in Singapore

To the memory of Ulrich Zähle

Preface

This book is based on a series of lectures at the Mathematics Department of the University of Jena developed in the period from 1995 up to 2015. It is completed by some additional material and extensions of known results from the literature to more general metric spaces.

Our aim is to give an introduction to classical fields of fractal geometry, which provide some background for modern topics of research and applications. Related monographs with references to the corresponding literature are [Edgar (1997)], [Falconer (1985)] and [Falconer (1997)] (with new editions), [Mattila (1995)], [Pesin (1997)], [Pesin and Climenhaga (2009)].

The present book is suited for Master and PhD students, but also for mathematicians from other fields interested in fractals. Some basic knowledge on general measure theory and on topological notions in metric spaces is presumed. (For courses the material can also be restricted to the Euclidean setting. Some of the exercises are included.)

In Chapter 1 we first recall the main notions and results of the latter used in the sequel. Then outer measures in metric spaces are studied in more detail, in particular a refinement of the Carathodory construction by means of optimal coverings.

In Chapter 2 the special case of Hausdorff measures with general gauge functions is investigated, followed by packing measures as a certain counterpart. The considered properties and relationships have been of continuous interest for known classes of fractals. In particular, related fractal dimension notions and relevant statements are derived.

In Chapter 3 upper and lower densities of Borel measures and their relationships to Hausdorff and packing measures are considered. In the next chapters they are used for estimating Hausdorff and packing dimensions. The so-called Frostman lemma is proved for general complete separable metric spaces, which seems to be an extension of known results.

Chapter 4 deals with a potential theoretic approach to lower estimates of Hausdorff dimensions. (In general, upper estimates are much easier.) In particular, it is shown that the Hausdorff dimension agrees with the capacity dimension. This method is frequently used for dimension estimates in certain physical and stochastic models. Examples for the latter are presented in Chapter 9.

Other fractal dimension notions, which are easier to handle, are introduced in Chapter 5. These are Minkowski and box dimensions in Euclidean spaces as well as entropy and metric dimensions in the general case. Moreover, their relationships between each other and to Hausdorff and packing dimensions are derived and illustrated by first examples.

In Chapter 6 local and global Hausdorff and packing dimensions of (finite) Borel measures are introduced. The latter are shown to coincide with the infimum of those for certain underlying sets. On the other side, the dimensions of Borel sets are expressed as the supremum of the dimensions of Borel measures concentrated on these sets.

The main Chapter 7 deals with attractors of contracting iterated function systems as one of the fundamental models of fractal geometry. Here methods from the previous chapters are used. For the special case of self-similar sets in Euclidean spaces we derive the well-known dimension results. Then invariant measures with weights in the general case are studied, in particular, their attractor and dimension properties. Moreover, for the Euclidean case average densities of self-similar measures are calculated. At the end we show an ergodic theorem which leads to a simple method for associated computer simulations.

Applications to the theory of dynamical systems are illustrated in Chapter 8. Here we consider the simple example of the logistic map with different parameters in full detail. This concerns the attractor with respect to the inverse branches being the repeller of the primary map, chaotic dynamics and dimension estimates.

In Chapter 9 the dimension problem for graphs of Hölder continuous functions and of self-similar stochastic processes is treated. For the latter some probabilistic background is supposed. Together with the potential theoretic approach from Chapter 5 we obtain, in particular, the well-known formula for the box and Hausdorff dimension of the graph of fractional Brownian motion.

M. Zähle

About the Author

Martina Zähle, being since 1988 Full Professor of Mathematics at the Friedrich Schiller University of Jena, retired in 2015. Up to now, she has been working on different scientific projects on fractals. She is the initiator and co-organiser of the conference series *Fractal Geometry and Stochastics I–V*. She is a member of the editorial boards of the *Journal of Fractal Geometry* and of the World Scientific book series *Fractals and Dynamics in Mathematics, Science, and the Arts*.

Acknowledgments

The author would like to thank all the participants, who checked parts of earlier versions of the manuscript over the years. This concerns, in particular, the former students Bettina Bühler, Jakob Creutzig and Benjamin Sambale, as well as Uta Freiberg, Michael Hinz and Steffen Winter who also held additional tutorials or seminars. My special thanks go to Marc Rauch for his careful reading, many remarks to the notes and new exercises during the last lecture period. I am also grateful to my son in law Christian, my daughter Iljana and their children Alina and Moritz, who produced the graphics in the last chapters.

Finally I am indebted to the related publishing team of World Scientific for the nice assistance, in particular, to Rochelle Kronzek for the long contact, to Ana Ovey and Aanand Jayaraman.

Contents

Chapter 1

Measure Theoretic Foundations

1.1 Measures and outer measures

At the beginning we will give a short survey on measure concepts in abstract spaces which are most relevant for our purposes.

Classical measure theory, as presented in graduate courses on analysis or probability theory, goes back to Borel, Lebesgue, Radon, Vitali, Young and others. It starts from contents and premeasures on semirings of sets and extends to σ-additive measures on generated σ-algebras. The latter are basic notions for the Lebesgue integration theory and probability spaces. In the same period a modified and, in a sense, more general approach was initiated by Carathéodory. His theory of outer measures has been continued by Hausdorff, Besicovitch, Federer, Rogers and others. It has had a great influence on the development of modern geometric measure theory and of fractal geometry. We now will recall some notions and results from both these lines.

Throughout the book we will use the set inclusion signs $\subset$ and $\supset$ also for $\subseteq$ and $\supseteq$, respectively. If necessary, the difference will be pointed out.

In this section the basic space X is an arbitrary set and $\mathfrak{P}(X)$ denotes the system of all subset of X. One main aim is to "measure" the elements of $\mathfrak{P}(X)$. As a primary model the volume (or content) concept of classical geometry is used.

Definition 1.1. A mapping $\mu : \mathfrak{P}(X) \to [0, \infty]$ is called *outer measure* if

  (i) $\mu(\emptyset) = 0$;
  (ii) $\mu(A) \leq \sum_{i=1}^{\infty} \mu(A_i)$, if $A \subset \bigcup_{i=1}^{\infty} A_i$ (σ-*subadditivity*).

Note that σ-subadditivity implies *monotonicity*, i.e., $\mu(A) \leq \mu(B)$ if $A \subset B$. In the sense of contents, however, this is not yet satisfactory, because for

1

disjoint sets A_i one would like to obtain equality in (ii). Therefore measure theoretic considerations are often restricted to "nice" systems of subsets of the basic space X.

Definition 1.2. A subsystem $\mathfrak{X}$ of $\mathfrak{P}(X)$ is called *σ-algebra* or *σ-field* over X if

(i) $X \in \mathfrak{X}$;
(ii) $A \in \mathfrak{X}$ implies $A^c := X \setminus A \in \mathfrak{X}$;
(iii) $A_1, A_2, \cdots \in \mathfrak{X}$ implies $\bigcup_{i=1}^{\infty} A_i \in \mathfrak{X}$.

This refines the notion of a set algebra, defined as a set ring containing the whole space. In the sequel $\mathfrak{X}$ is always a σ-algebra on X.

Exercise 1.1. Show that

(i) $\emptyset \in \mathfrak{X}$ and $\bigcap_{i=1}^{\infty} A_i \in \mathfrak{X}$ for $A_1, A_2, \cdots \in \mathfrak{X}$,
(ii) in Definition 1.2 (iii) the union $\bigcup_{i=1}^{\infty}$ may be replaced by the intersection $\bigcap_{i=1}^{\infty}$.

Definition 1.3. The pair $[X, \mathfrak{X}]$ is called *measurable space*.

σ-algebras, and hence measurable spaces, generated by a system $\mathcal{E}$ of subsets of X play a crucial role in many considerations.

Definition 1.4. The smallest σ-algebra $\sigma(\mathcal{E})$ on X containing $\mathcal{E}$ is said to be *generated by the set system* $\mathcal{E}$.

Exercise 1.2. Show that
$$\sigma(\mathcal{E}) = \bigcap \{\mathfrak{X} : \mathfrak{X} \text{ is a σ-algebra on } X \text{ containing } \mathcal{E}\}.$$

(We use the symbol $\bigcap$ also for intersections of systems.)

Definition 1.5. A *measure* on the measurable space $[X, \mathfrak{X}]$ (or on $\mathfrak{X}$) is a mapping $\mu : \mathfrak{X} \to [0, \infty]$ such that

(i) $\mu(\emptyset) = 0$,
(ii) $\mu(\bigcup_{i=1}^{\infty} A_i) = \sum_{i=1}^{\infty} \mu(A_i)$, if the sets $A_1, A_2, \ldots \in \mathfrak{X}$ are pairwise disjoint (*σ-additivity*).

Exercise 1.3. Any measure μ on $[X, \mathfrak{X}]$ may be extended to an outer measure μ^* on $[X, \mathfrak{P}(X)]$ by
$$\mu^*(A) = \inf\{\mu(B) : B \in \mathfrak{X}, \ A \subset B\}.$$
(Below we will recall the opposite approach, i.e., restricting outer measures to measurable sets in order to obtain measures.)

Definition 1.6. A triple $[X, \mathfrak{X}, \mu]$ as above is called *measure space*.

Remark 1.7. In the case of $\mu(X) = 1$ one also speaks of a *probability measure* and a *probability space*, respectively.

Theorem 1.8 (Continuity of measures). *Let $[X, \mathfrak{X}, \mu]$ be a measure space and $A_1, A_2, \ldots \in \mathfrak{X}$. Then we have*

(i) $\mu\left(\bigcup_{n=1}^{\infty} A_n \right) = \lim\limits_{n \to \infty} \mu(A_n), \quad if \quad A_1 \subset A_2 \subset \cdots ,$

(ii) $\mu\left(\bigcap_{n=1}^{\infty} A_n \right) = \lim\limits_{n \to \infty} \mu(A_n), \quad if \quad A_1 \supset A_2 \supset \cdots \quad and \quad \mu(A_1) < \infty ,$

(iii) $\mu\left(\bigcup_{k=1}^{\infty} \bigcap_{n=k}^{\infty} A_n \right) \leq \liminf\limits_{n \to \infty} \mu(A_n) .$

These notions and the resulting theory are usually presented in basic courses of measure theory. We refer, e.g., to [Bauer (2001)], [Cohn (2013)], [Elstrodt (1996)] and [Halmos (1974)].

In the sequel we will adopt Carathéodory's point of view with outer measures as the primary subject of study. Here we also refer to [Rogers (1998)], [Mattila (1995)] and [Federer (1969)]. From now on let μ be an outer measure on X.

Definition 1.9. $A \subset X$ is said to be *μ-measurable*, if for all $E \subset X$ we have

$$\mu(E) = \mu(E \cap A) + \mu(E \setminus A) .$$

The family of μ-measurable subsets of X will be denoted by $\mathfrak{X}_\mu$. It provides a natural relationship to the above concepts:

Theorem 1.10.

(i) $\mathfrak{X}_\mu$ *is a σ-algebra on X.*

(ii) μ *regarded on $[X, \mathfrak{X}_\mu]$ is a measure.*

The proof is left to the reader as an exercise. It may be found in the monographs mentioned above.

We next will describe two methods for generating outer measures. Throughout the book for $A \subset X$ the *restriction of μ to A* will be denoted by $\mu \llcorner A$, i.e,

$$\mu \llcorner A(B) := \mu(A \cap B), \quad B \subset X.$$

Exercise 1.4. Show that

    (i) $\mu \llcorner A$ is again an outer measure on X;

    (ii) $\mathfrak{X}_\mu \subset \mathfrak{X}_{\mu \llcorner A}$.

(We use "$\subset$" also for inclusion of systems.)

From 1.3 we know that every measure on a σ-algebra of X induces an outer measure. The following provides a general method of extending simple set functions to outer measures on X.

Definition 1.11. Let $\mathcal{C}$ be an arbitrary family of subsets of X such that $\emptyset \in \mathcal{C}$. A mapping $\tau : \mathcal{C} \to [0, \infty]$ will be called *weight function* on $\mathcal{C}$ if $\tau(\emptyset) = 0$.

Note that this abstract notion includes that of premeasures on semirings of sets, which are introduced as σ-additive contents. Then *Carathéodory's construction* is as follows:

Theorem 1.12. *For $\mathcal{C}$ and τ as in Definition 1.11, the set function*

$$\mu(A) := \inf \left\{ \sum_{i=1}^{\infty} \tau(C_i) : C_i \in \mathcal{C}, \bigcup_{i=1}^{\infty} C_i \supset A \right\}$$

is an outer measure on X.

(Cf. [Rogers (1998)] or [Elstrodt (1996)].) This construction is consistent with the interpretation of μ as a weight function in the above sense:

Exercise 1.5.

    (i) Show that any outer measure μ agrees with the result of the previous construction for $\mathcal{C} := \mathfrak{P}(X)$ and $\tau(C) := \mu(C)$, $C \in \mathcal{C}$.

    (ii) Let ν be a measure on $[X, \mathfrak{X}]$ and ν^* be the induced outer measure from 1.3. Set $\mathcal{C} := \mathfrak{X}$ and $\tau(C) := \nu(C)$ for $C \in \mathcal{C}$ and define μ as before. Then we get $\mu = \nu^*$.

Remark 1.13. Authors working in geometric measure theory frequently replace the name "outer measure" by "measure" and "measure" by "σ-additive measure" on a σ-algebra. When it is clear from the context we will also follow such a convention.

1.2 Outer measures in metric spaces

In the case of metric spaces (X, d) the construction of certain outer measures and the study of their properties is simplified by the structure of the space.

Definition 1.14. The *Borel σ-algebra* $\mathfrak{B} = \mathfrak{B}(X)$ on (X, d) is the smallest σ-algebra containing all open sets (i.e., the σ-algebra generated by the open sets).

Exercise 1.6. $\mathfrak{B}$ is also generated by each of the following families:

(i) the closed sets;
(ii) the open (or closed) balls if X is separable;
(iii) the n-dimensional open (closed or half-closed) rectangles if X is the Euclidean space $\mathbb{R}^n$.

Definition 1.15. An outer measure μ on (X, d) is called *Borel (outer) measure* if all Borel sets are μ-measurable (briefly, if $\mathfrak{B} \subset \mathfrak{X}_\mu$).

The following notion of metric measures provides a nice equivalent condition for the Borel property. Throughout the book we will use the notations
$$d(A, B) := \inf\{d(a, b) : a \in A, \ b \in B\}$$
and $d(x, B) := d(\{x\}, B)$ for $A, B \subset X$, $x \in X$.

Definition 1.16. An outer measure μ is said to be *metric* if
$$\mu(A \cup B) = \mu(A) + \mu(B)$$
for any $A, B \subset X$ such that $d(A, B) > 0$.

(In case of positive pseudodistance $d(A, B)$ the sets A and B are said to be *positively separated.*)

Theorem 1.17 (Carathéodory). *An outer measure on a metric space is a Borel measure if and only if it is metric.*

Proof. For the 'if part' it suffices to prove that all closed sets are μ-measurable by the following three arguments: $\mathfrak{B}$ is generated by these sets; $\mathfrak{X}_\mu$ is a σ-algebra; any σ-algebra contains together with a system $\mathcal{E}$ the σ-algebra generated by $\mathcal{E}$. Thus it is enough to show for any closed set C that
$$\mu(E) \geq \mu(E \cap C) + \mu(E \setminus C), \qquad E \subset X,$$
since the opposite inequality follows from the subadditivity of μ.

The case $\mu(E) = \infty$ is trivial. Assume $\mu(E) < \infty$ and consider the auxiliary parallel sets

$$C_i = \{x \in X : d(\{x\}, C) \leq \frac{1}{i}\}, \qquad i = 1, 2, \ldots .$$

Then the sets $E \cap C$ and $E \setminus C_i$ are positively separated. Hence,

$$\mu(E) \geq \mu((E \cap C) \cup (E \setminus C_i)) = \mu(E \cap C) + \mu(E \setminus C_i).$$

The proof is complete if we show that

$$\lim_{i \to \infty} \mu(E \setminus C_i) = \mu(E \setminus C).$$

Since C is closed, we can use the representation

$$E \setminus C = (E \setminus C_i) \cup \bigcup_{k=i}^{\infty} D_k, \text{ with } D_k := (C_{k-1} \setminus C_k) \cap E$$

(for $i \geq 2$) and obtain

$$\mu(E \setminus C_i) \leq \mu(E \setminus C) \leq \mu(E \setminus C_i) + \sum_{k=i}^{\infty} \mu(D_k).$$

Therefore it remains to show that $\lim_{i \to \infty} \sum_{k=i}^{\infty} \mu(D_k) = 0$, which is equivalent to $\sum_{k=2}^{\infty} \mu(D_k) < \infty$.

From $d(D_j, D_k) > 0$, if $|j - k| \geq 2$, and the metric property of μ we infer

$$\sum_{l=2}^{n} \mu(D_{2l}) = \mu\left(\bigcup_{l=2}^{n} D_{2l}\right) \leq \mu(E), \quad \sum_{l=2}^{n} \mu(D_{2l-1}) = \mu\left(\bigcup_{l=2}^{n} D_{2l-1}\right) \leq \mu(E).$$

This yields $\sum_{i=2}^{n} \mu(D_i) \leq 2\mu(E)$ and consequently, $\sum_{k=2}^{\infty} \mu(D_k) < \infty$.

For the 'only if' part it suffices to prove

$$\mu(A \cup B) \geq \mu(A) + \mu(B)$$

for positively separated sets, since the opposite inequality follows from sub-additivity. The above inequality is a consequence of the representation $\mu(C) = \inf\{\mu(\tilde{C}) : \tilde{C} \in \mathfrak{B}, \ C \subset \tilde{C}\}$ (cf. Exercise 1.5, (ii)). The details are left to the reader as a new exercise. (See also [Federer (1969)][p. 74-75].) □

A classical example is given by the *outer Lebesgue measure* in $\mathbb{R}^n$: Let $\mathcal{R}$ be the system of open n-dimensional rectangles in $\mathbb{R}^n$. For $R = (a_1, b_1) \times \cdots \times (a_n, b_n) \in \mathcal{R}$ the *n-volume (content)* is determined by $V^n(R) := \prod_{i=1}^{n}(b_i - a_i)$. The outer Lebesgue measure $\mathcal{L}^n$ arises by the Carathéodory method described in Theorem 1.12 for the weight function $\tau(R) := V^n(R)$ on $\mathcal{R}$:

Definition 1.18. For $A \subset \mathbb{R}^n$ let

$$\mathcal{L}^n(A) := \inf\left\{\sum_{i=1}^{\infty} V^n(R_i) : R_i \in \mathcal{R}, \ A \subset \bigcup_{i=1}^{\infty} R_i\right\}.$$

Theorem 1.19. $\mathcal{L}^n$ *is a metric outer measure on* $\mathbb{R}^n$ *with the property* $\mathcal{L}^n(R) = V^n(R)$ *for any rectangle* $R \in \mathcal{R}$.

Proof. The outer measure property is a special case of Theorem 1.12. By construction, $\mathcal{L}^n(R) \leq V^n(R)$. In order to prove the opposite inequality for $R = \times_{i=1}^n (a_i, b_i)$ we consider the closed n-rectangles $R_\varepsilon := \times_{k=1}^n [a_k + \varepsilon, b_k - \varepsilon]$. Monotonicity of $\mathcal{L}^n$ implies

$$\lim_{\varepsilon \to 0} \mathcal{L}^n(R_\varepsilon) \leq \mathcal{L}^n(R).$$

Therefore it is enough to prove $\prod_{k=1}^n (d_k - c_k) \leq \mathcal{L}^n(\overline{R})$ for any closed rectangle $\overline{R} = \times_{k=1}^n [c_k, d_k]$. By the Heine-Borel lemma any covering of $\overline{R}$ by open rectangles contains a finite subcovering.

Thus, it remains to show

$$\prod_{k=1}^n (d_k - c_k) \leq \sum_{i=1}^N V^n(R_i), \quad \text{if} \quad R_i \in \mathcal{R} \quad \text{and} \quad \overline{R} \subset \bigcup_{i=1}^N R_i.$$

This is easy to see when subdividing the intervals (a_k, b_k) by means of the endpoints of the corresponding intervals of R_i, $i = 1, \ldots, N$.

Finally, we have to prove that $\mathcal{L}^n$ is metric. This is also a consequence of Carathéodory's extension Theorem 1.12, since for positively separated sets the covering rectangle systems may be chosen disjoint. $\square$

Turning back to general Borel measures, we now introduce an additional condition:

Definition 1.20. An outer measure μ on a metric space X is said to be *Borel regular*, if it is a Borel measure and for any $A \subset X$ there exists a Borel set $B \supset A$ with $\mu(B) = \mu(A)$.

Proposition 1.21. *If* μ *is a Borel regular measure,* A *is a* μ-*measurable set and* $\mu(A) < \infty$, *then* $\mu \llcorner A$ *is also a Borel regular measure.*

Proof. Choose $B \in \mathfrak{B}$ such that $\mu(B) = \mu(A)$. Then we have $\mu(B \setminus A) = 0$. Let now $C \subset X$ be an arbitrary set. Since μ is Borel regular, there exists a Borel set D such that $B \cap C \subset D$ and $\mu(B \cap C) = \mu(D)$. Note that $C \subset D \cup B^c =: E \in \mathfrak{B}$. We will show that E is the required Borel set for C with respect to $\mu \llcorner A$. By construction we get $B \cap E = B \cap D$. Moreover we have $A \cap E \subset D$ and thus,

$$\mu(A \cap E) \leq \mu(D) \leq \mu(B \cap C) \leq \mu(A \cap C) + \mu(B \setminus A).$$

Since $\mu(B \setminus A) = 0$, this implies $\mu \llcorner A(E) \leq \mu \llcorner (C)$, and hence equality. $\square$

Approximation from above by means of open sets (or of Borel sets) is a sufficient condition for the regularity of a Borel measure:

Proposition 1.22. *If μ is a Borel measure on the metric space X satisfying*

$$\mu(A) = \inf\{\mu(G) : G \text{ open}, A \subset G\}$$

$$(\text{or } \mu(A) = \inf\{\mu(B) : B \in \mathfrak{B}, A \subset B\})$$

for any $A \subset X$, then μ is Borel regular.

Proof. We consider the case of approximation by open sets. For Borel sets the arguments are analogous. If $\mu(A) = \infty$ then $\mu(A) = \mu(X)$. Thus, X is the desired Borel set. If $\mu(A) < \infty$ the definition of the infimum implies the existence of a decreasing sequence of open set G_n containing A such that $\mu(G_n) < \mu(A) + 1/n$. Putting $B := \bigcap_{n=1}^{\infty} G_n$ we obtain that $B \in \mathfrak{B}$, $A \subset B$ and, by Theorem 1.8 (ii),

$$\mu(B) = \lim_{n \to \infty} \mu(G_n) \leq \mu(A).$$

Hence, $\mu(B) = \mu(A)$. □

Under certain finiteness assumption this condition is also necessary for Borel regularity. Furthermore, for the case of Borel regular measures and measurable sets with finite measure one also gets approximation from below by closed sets.

Definition 1.23. In the sequel an outer measure μ on X is said to be σ-*finite*, if there is an increasing sequence of open sets X_n with $\bigcup_{n=1}^{\infty} X_n = X$ such that $\mu(X_n) < \infty$ for all n.

(Usually in measure theory the X_n in the definition of σ-finiteness are assumed to be only Borel sets, not necessarily open. For our purposes the stronger condition is essential. Most often we will work with Borel measures which are finite at least on closed balls.)

Theorem 1.24. *Let μ be an outer Borel measure on a metric space X.*

(i) *Suppose that μ is σ-finite. If $A \in \mathfrak{B}$, or $A \subset X$ and μ is Borel regular, then we have*

$$\mu(A) = \inf\{\mu(G) : G \text{ open}, A \subset G\}.$$

(ii) *Suppose that $\mu(A) < \infty$. If $A \in \mathfrak{B}$, or μ is Borel regular and $A \in \mathfrak{X}_\mu$, then we have*

$$\mu(A) = \sup\{\mu(F) : F \text{ closed}, F \subset A\}.$$

Proof. We will proceed in three steps.

Step 1. Choose the sets X_n as in Definition 1.23. Let $\mathfrak{B}_n$ be the system of Borel subsets B of X_n such that for any $\varepsilon > 0$ there are an open set $G \supset B$ with $\mu(G \setminus B) < \varepsilon$ and a closed set $F \subset B$ with $\mu(B \setminus F) < \varepsilon$. Then it is straightforward to prove that $\mathfrak{B}_n$ is a σ-algebra on X_n. It contains all closed subsets of X_n, since μ is continuous as measure on the Borel sets (Theorem 1.8), and any closed set may be represented as intersection of a decreasing sequence of open set. Therefore $\mathfrak{B}_n$ agrees with the Borel σ-algebra on X_n. This means that assertions (i) and (ii) hold true for all Borel sets in X_n, since $0 \le \mu(G) - \mu(B) = \mu(G \setminus B)$ and $0 \le \mu(B) - \mu(F) = \mu(B \setminus F)$.

Step 2. We will derive assertion (i) for arbitrary $A \subset X$. Choose $B \in \mathfrak{B}$ such that $A \subset B$ and $\mu(A) = \mu(B)$. (If A is a Borel set, we can take $B := A$.) Using Step 1 for the X_n as above one obtains

$$\mu(A) \le \inf\{\mu(G) : G \text{ open}, \ A \subset G\}$$
$$\le \inf\{\mu(G) : G \text{ open}, \ B \subset G\}$$
$$\le \liminf_{n \to \infty}\{\mu(G) : G \text{ open}, \ B \cap X_n \subset G\}$$
$$= \lim_{n \to \infty} \mu(B \cap X_n) = \mu(B) = \mu(A),$$

i.e, the claimed equality.

Step 3. In order to infer (ii) for $A \in \mathfrak{X}_\mu$ with $\mu(A) < \infty$ we first will show that A contains a Borel set D with the same measure. Since μ is Borel-regular, A is contained in a Borel set B with $\mu(A) = \mu(B)$. From $\mu(A) < \infty$ we infer

$$\mu(B \setminus A) = \mu(B) - \mu(A) = 0.$$

By the Borel-regularity of μ there is a $C \in \mathfrak{B}$ such that $B \setminus A \subset C$ and

$$\mu(C) = \mu(B \setminus A) = 0.$$

Then $D := B \setminus C$ is the required Borel set: Obviously, $D \subset A$. The inclusion $A \setminus D \subset C$ implies $\mu(A \setminus D) = 0$ and, by measurability of A, $\mu(A) = \mu(D)$.

Therefore it is enough to prove (ii) for arbitrary $B \in \mathfrak{B}$. This is again a consequence of Step 1 when considering the finite measure $\mu \llcorner A$ instead of μ. $\square$

In the sequel we will also need the following *approximation from below by compact sets*:

Theorem 1.25. *Let X be a complete separable metric space and $\mu(A) < \infty$, where μ is a Borel regular measure on X and $A \subset X$, or μ is a Borel measure on X and $A \in \mathfrak{B}$. Then we have*

$$\mu(A) = \sup\{\mu(K) : K \text{ compact}, \ K \subset A\}.$$

Proof. In view of Theorem 1.24 for any $\epsilon > 0$ there exists a closed set $F \subset A$ such that $\mu(A \setminus F) < \varepsilon/2$. Now let $\{x_k : k \in \mathbb{R}^n\}$ be a dense subset of X and $B(x_k, 2^{-n})$ be the closed ball with centre at x_k and radius 2^{-n}. Denote $F_{k,n} := F \cap B(x_k, 2^{-n})$. For any n these sets form a covering of F, hence there exists a number N_n such that

$$\mu(F) - \mu\Big(\bigcup_{k=1}^{N_n} F_{k,n}\Big) < \frac{\varepsilon}{2^{n+1}}.$$

Set

$$K := \bigcap_{n=1}^{\infty} \bigcup_{k=1}^{N_n} F_{k,n}.$$

K is closed and admits, for any n, a finite 2^{-n}-covering, i.e., K is totally bounded. Since X is complete, K is compact. Furthermore,

$$\mu(A \setminus K) = \mu(A \setminus F) + \mu(F \setminus K) \leq \frac{\varepsilon}{2} + \sum_{n=1}^{\infty} \mu\Big(F \setminus \bigcup_{k=1}^{N_n} F_{k,n}\Big)$$

$$= \frac{\varepsilon}{2} + \sum_{n=1}^{\infty} \Big(\mu(F) - \mu\big(\bigcup_{k=1}^{N_n} F_{k,n}\big)\Big) \leq \frac{\varepsilon}{2} + \sum_{n=1}^{\infty} \frac{\varepsilon}{2^{n+1}} = \varepsilon.$$

Hence, for any $\varepsilon > 0$, we have found a compact set $K \subset A$ such that $\mu(A \setminus K) \leq \varepsilon$. This leads to the assertion. $\square$

Finally, we will describe a *refinement of Carathéodory's construction* 1.12 for outer measures in the case of metric spaces, which leads to Borel measures. Some natural specifications, e.g. Hausdorff measures, provide also Borel regularity. The idea is to consider finer and finer coverings of a set.

By $|C|$ we will denote the *diameter* of a set C in the metric space (X, d), i.e.,

$$|C| := \sup\{d(x, y) : x, y \in C\}.$$

Let $\mathcal{C}$ again be an arbitrary system of subsets of X and $\delta > 0$.

Definition 1.26. A sequence of sets $(C_i)_{i=1}^{\infty}$ is called a $(\mathcal{C}, \delta) - covering$ of C, if $C_i \in \mathcal{C}$, $|C_i| \leq \delta$ for all i and $A \subset \bigcup_{i=1}^{\infty} C_i$. If $\mathcal{C} = \mathfrak{P}(X)$ one speaks of a δ-covering.

For a given weight function τ on $\mathcal{C}$, we may introduce the outer measure with respect to the subsystem $\mathcal{C}_\delta$ of sets with diameter less than δ, according to Theorem 1.12:

Corollary 1.27. *For any $\delta > 0$ the set function*

$$\mu_\delta^\tau(A) = \mu_\delta^{\tau,\mathcal{C}}(A) := \inf \left\{ \sum_{i=1}^\infty \tau(C_i) : (C_i)_{i=1}^\infty \text{ is a } (\mathcal{C}, \delta) - \text{covering of } A \right\}$$

determines an outer measure on X.

In general, μ_δ^τ is poor of measurable sets.

Exercise 1.7. Construct an example for (X, d), $\mathcal{C}$, τ, δ, a Borel set B and a set $E \subset X$ such that

$$\mu_\delta^\tau(E) < \mu_\delta^\tau(E \cap B) + \mu_\delta^\tau(E \setminus B).$$

The situation becomes better if we take the limit as $\delta \to 0$. This is possible, since

$$\mu_\delta^\tau(A) \geq \mu_{\delta'}^\tau(A) \quad \text{if} \quad \delta < \delta', \quad A \subset X.$$

Denote

$$\mu^\tau(A) = \mu^{\tau,\mathcal{C}}(A) := \lim_{\delta \to 0} \mu_\delta^\tau(A) = \sup_{\delta > 0} \mu_\delta^\tau(A).$$

Proposition 1.28. μ^τ *is a metric outer measure on X.*

Proof. By the σ-subadditivity of μ_δ^τ and the monotone convergence we get for any $A \subset \bigcup_{i=1}^\infty A_i$,

$$\mu^\tau(A) \leq \lim_{\delta \to 0} \sum_{i=1}^\infty \mu_\delta^\tau(A_i) \leq \sum_{i=1}^\infty \lim_{\delta \to 0} \mu_\delta^\tau(A_i) = \sum_{i=1}^\infty \mu^\tau(A_i).$$

In order to show that μ^τ is metric (recall Definition 1.16), suppose $d(A, B) > 0$, for $A, B \subset X$. Choose $\delta < d(A, B)/2$. Then any $(\mathcal{C}, \delta)$-covering of $A \cup B$ provides two disjoint $(\mathcal{C}, \delta)$-coverings of A and B, respectively. Hence,

$$\mu^\tau(A \cup B) \geq \mu^\tau(A) + \mu^\tau(B),$$

but this yields equality because of the subadditivity of μ^τ. □

As a consequence of Theorem 1.17 and Proposition 1.28, μ^τ is a Borel measure. In the next section we will specify this construction to Hausdorff measures.

Chapter 2

Hausdorff and Packing Measures

2.1 Hausdorff measures

Continuing the last considerations we will now choose $\mathcal{C} := \mathfrak{P}(X)$ as covering system on the metric space (X, d). The weight function τ will be given by means of auxiliary functions defined on the diameters of the sets. Denote

$$\mathbb{H} := \{h : [0, \infty] \to [0, \infty] : h(u) > 0, \text{ for } u > 0, \tag{2.1}$$

$$h \text{ is right continuous and non-decreasing}\}. \tag{2.2}$$

$\mathbb{H}$ is called the system of *gauge functions* h for the construction. Important examples are

$$h(u) := u^s \text{ or } h(u) := u^s (\log |\log u|)^t \text{ for } u \le \varepsilon \text{ and } h(u) := h(\varepsilon) \text{ otherwise}$$

(with ε sufficiently small, $s \ge 0$, and $t \ge 0$).

Definition 2.1. Let $h \in \mathbb{H}$, $\mathcal{C} := \mathfrak{P}(X)$ and the weight function τ be given by $\tau(C) := h(|C|)$ if $C \ne \emptyset$. Then the outer measure μ^τ arising in Carathéodory's construction by means of finer and finer coverings (see Corollary 1.27 and Proposition 1.28) is called *Hausdorff measure corresponding to the gauge function* h or, briefly, *Hausdorff h-measure*. It is denoted by $\mathcal{H}^h$, i.e.,

$$\mathcal{H}^h(A) := \lim_{\delta \to 0} \mathcal{H}^h_\delta(A), \text{ where}$$

$$\mathcal{H}^h_\delta(A) := \inf \left\{ \sum_{i=1}^\infty h(|C_i|) : \{C_i\}_{i=1}^\infty \text{ is a } \delta\text{-covering of } A \right\}.$$

In the special case $h(u) = u^s$ one writes $\mathcal{H}^s := \mathcal{H}^h$ and calls this *s-dimensional Hausdorff measure*. Similarly, $\mathcal{H}^s_\delta$ denotes the corresponding δ-approximating (outer) measure.

(The ideas for the above notions go back to [Hausdorff (1919)].) Note that for $s = 0$ one obtains the counting measure $\mathcal{H}^0$ determined by the number of elements, if the set is finite, and being infinite otherwise.

In order to prove the Borel regularity of Hausdorff measures we first show that the covering system may be reduced to topologically nice sets: By $\overline{C}$ we denote the closure of a set $C \subset X$.

Theorem 2.2. *In the definition of the Hausdorff measure $\mathcal{H}^h$ the covering system $\mathfrak{P}(X)$ may be replaced by each of the following ones:*

    (i) *the system $\mathcal{F}$ of all closed subsets of the metric space X,*
    (ii) *the system $\mathcal{G}$ of all open subsets of X.*

Proof. (i) Since $|\overline{C}| = |C|$, together with $(C_i)_{i=1}^{\infty}$ the sequence $(\overline{C}_i)_{i=1}^{\infty}$ is also a δ-covering of $A \subset X$. Hence,

$$\mathcal{H}_{\delta}^{h,\mathcal{F}}(A) \leq \mathcal{H}_{\delta}^h(A).$$

The opposite inequality is obvious. Hence, we infer

$$\mathcal{H}^{h,\mathcal{F}}(A) = \lim_{\delta \to 0} \mathcal{H}_{\delta}^{h,\mathcal{F}}(A) = \lim_{\delta \to 0} \mathcal{H}_{\delta}^h(A) = \mathcal{H}^h(A).$$

(ii) If we can show that

$$\mathcal{H}_{\delta}^{h,\mathcal{G}}(A) \leq \mathcal{H}_{\delta'}^h(A), \tag{2.3}$$

for any $\delta > \delta'$, then the proof is complete, since this implies $\mathcal{H}^{h,\mathcal{G}}(A) \leq \mathcal{H}^h(A)$, letting $\delta \to 0$. Together with the obvious inequality $\mathcal{H}^h(A) \leq \mathcal{H}^{h,\mathcal{G}}(A)$ we then have equality.

It suffices to consider the case $\mathcal{H}_{\delta'}^h(A) < \infty$. For any $\varepsilon > 0$ we can choose a δ'-covering $(C_i)_{i=1}^{\infty}$ of A such that

$$\sum_{i=1}^{\infty} h(|C_i|) < \mathcal{H}_{\delta'}^h(A) + \varepsilon. \tag{2.4}$$

Then we set $G_i := \emptyset$, if $C_i = \emptyset$, and construct for $C_i \neq \emptyset$ an open set $G_i \supset C_i$ with $|G_i| < \delta$: Right continuity of h at $u := |C_i|$ yields

$$h(|C_i| + 2\varepsilon_i) < h(|C_i|) + \frac{\varepsilon}{2^i}, \tag{2.5}$$

for all sufficiently small ε_i satisfying $\delta' + 2\varepsilon_i < \delta$.

Consider now the open parallel set

$$G_i = \{x \in X : d(\{x\}, C_i) < \varepsilon_i\}.$$

Monotonicity of h together with $|G_i| \leq |C_i| + 2\varepsilon_i$ and (2.5) imply

$$h(|G_i|) \leq h(|C_i| + 2\varepsilon_i) < h(|C_i|) + \frac{\varepsilon}{2^i}.$$

Therefore,

$$\sum_{i=1}^{\infty} h(|G_i|) \le \sum_{i=1}^{\infty} h(|C_i|) + \varepsilon < \mathcal{H}_{\delta'}^{h}(A) + 2\varepsilon$$

in view of (2.4). Hence,

$$\mathcal{H}_{\delta}^{h,\mathcal{G}} < \mathcal{H}_{\delta'}^{h}(A) + 2\varepsilon \,.$$

Letting $\varepsilon \to 0$ we obtain the remaining inequality (2.3). $\qquad\square$

Borel-regularity is now a consequence:

Corollary 2.3. $\mathcal{H}^h$ *is Borel regular.*

Proof. By Theorem 2.2 there exists a sequence of open $1/n$-coverings $(G_i^n)_{i=1}^{\infty}$ of $A \subset X$ such that

$$\sum_{i=1}^{\infty} h(|G_i^n|) < \mathcal{H}_{1/n}^{h}(A) + \frac{1}{n} \,,$$

$n = 1, 2, \ldots$, provided that $\mathcal{H}^h(A) < \infty$. Set $B := \bigcap_{n=1}^{\infty} \bigcup_{i=1}^{\infty} G_i^{(n)}$. Then B is a Borel set containing A. Furthermore,

$$\mathcal{H}_{1/n}^{h}(B) \le \sum_{i=1}^{\infty} h(|G_i^n|) < \mathcal{H}_{1/n}^{h}(A) + \frac{1}{n} \,.$$

Letting $n \to \infty$ we infer $\mathcal{H}^h(B) \le \mathcal{H}^h(A)$. Hence, the inclusion $A \subset B$ yields $\mathcal{H}^h(B) = \mathcal{H}^h(A)$. The remaining case $\mathcal{H}^h(A) = \infty$ is trivial. $\qquad\square$

Remark 2.4. If the covering system $\mathcal{C}$ is further restricted to the system of all closed balls in X, then the resulting outer measure, denoted by $\mathcal{S}^h$, is called *spherical measure* corresponding to the gauge function h. For $h(u) = u^s$ we denote $\mathcal{S}^s := \mathcal{S}^h$. Similarly to Theorem 2.2 one can show that in the Euclidean case the closed balls may be replaced by the open balls. The Borel regularity of $\mathcal{S}^h$ follows as in 2.3. Note that, in general,

$$\mathcal{H}^h(A) \le \mathcal{S}^h(A).$$

Definition 2.5. A gauge function $h \in \mathbb{H}$ satisfies the *doubling condition* if for some constant $K > 0$,

$$h(2u) \le Kh(u), \quad u > 0.$$

Every $C \subset X$ with $|C| < \delta$ is contained in a ball of radius δ. Therefore $\mathcal{S}_{2\delta}^h(A) \leq K\mathcal{H}_\delta^h(A)$ and hence,

$$\mathcal{S}^h(A) \leq K\mathcal{H}^h(A)$$

provided the doubling condition is fulfilled.

For $h(u) = u^s$ the optimal constant in the doubling condition equals $K = 2^s$. In the Euclidean case $X = \mathbb{R}^n$ *for integer s* the last inequality may be improved by $\mathcal{S}^s(A) \leq \left(\frac{2n}{n+1}\right)^{s/2}\mathcal{H}^s(A)$ (see [Federer (1969), 2.10.42]).

In Section 2.2 we will show that $\mathcal{H}^n = \mathcal{S}^n = \mathrm{const}\,\mathcal{L}^n$ if $X = \mathbb{R}^n$. If we neglect multiplicative constants, this assertion is an easy consequence of the following *invariance properties.*

For a mapping $f : X \to X'$ and an outer measure μ on X we write $\mu \circ f^{-1}(A') := \mu\big(f^{-1}(A')\big)$ for the image measure in X'. In case $X = X'$, μ is said to be *f-invariant* if $\mu \circ f^{-1} = \mu$.

Definition 2.6. A *similarity* with *ratio* $r > 0$ is a mapping $f : (X, d) \to (X', d')$ with $d'(f(x), f(y)) = r\,d(x, y)$, $x, y \in X$. For $r = 1$ the mapping f is called an *isometry*.

Note that any similarity $f : X \to X'$ is a bijection from X onto $f(X) \subset X'$.

Proposition 2.7.

   (i) $\mathcal{H}^s\llcorner(f(X)) = r^s\,\mathcal{H}^s \circ f^{-1}$, $s \geq 0$, *for any similarity $f : (X, d) \to (X', d')$ with ratio r.*

   (ii) $\mathcal{H}^h\llcorner(f(X)) = \mathcal{H}^h \circ f^{-1}$, $h \in \mathbb{H}$, *for any isometry $f : (X, d) \to (X', d')$.*

   (iii) (i) *and* (ii) *remain valid for the outer measures $\mathcal{S}^h$ and $\mathcal{L}^n$. In particular, these measures are invariant under surjective isometries on the metric spaces X and $\mathbb{R}^n$, respectively.*

Remark 2.8. Recall that isometries in $\mathbb{R}^n$ are always surjective and may be represented as composition of an orthogonal linear mapping and a translation. Similarities in $\mathbb{R}^n$ are compositions of isometries and *homotheties* $f(x) = rx$ with ratios $r > 0$.

By 2.7, $\mathcal{L}^n$, $\mathcal{H}^n$ and $\mathcal{S}^n$ are all translation invariant measures on $\mathfrak{B}(\mathbb{R}^n)$. By construction they are σ-finite. Using Fubini's theorem one can easily show that they must be constant multiples one of each other (cf. uniqueness of Haar measure in the literature).

Proof of Proposition 2.7. Suppose $A' \subset f(X)$. Then the sets $(C_i)_{i=1}^{\infty}$ form a δ-covering of $A := f^{-1}(A')$ if and only if the sets $(f(C_i))_{i=1}^{\infty}$ form an $r\delta$-covering of A', since $|f(C)| = r|C|$. Hence,

$$\mathcal{H}_{r\delta}^s(A') = r^s \mathcal{H}_{\delta}^s(A) , \quad s \geq 0,$$

and for isometries ($r = 1$)

$$\mathcal{H}_{\delta}^h(A') = \mathcal{H}_{\delta}^h(A) , \quad h \in \mathbb{H}.$$

Letting $\delta \to 0$ we obtain the assertion for the Hausdorff measure. The arguments for the spherical measure and the Lebesgue measure are similar. □

The interpretation of $\mathcal{H}^n$ as the Lebesgue measure in $\mathbb{R}^n$ and the local chart approach of differential geometry lead to a relationship between Hausdorff measure of integer dimension and classical calculus:

Remark 2.9. If M_m is an m-dimensional smooth submanifold of $\mathbb{R}^n$, then $\mathcal{H}^m \llcorner M_m$ as a measure on the Borel σ-algebra agrees with the usual *surface area measure*. (Proofs and extensions to rectifiable sets may be found in [Federer (1969)]. See also [Mattila (1995)].)

Lipschitz or *Hölder mappings* play a central role in geometric measure theory, in particular in the situation sketched in the last remark.

Definition 2.10. $f : (X, d) \to (X', d')$ is said to be (*locally uniformly*) *Hölder continuous of order* $0 < \alpha \leq 1$ if

$$d'(f(x), f(y)) \leq \lambda d(x, y)^{\alpha} , \quad x, y \in X \quad (\text{and } d(x, Y) < c), \text{ resp.})$$

for some $\lambda > 0$ (and some $c > 0$, resp.).

For $\alpha = 1$ we will speak of *Lipschitz mappings*.

(Frequently Hölder functions are also called Lipschitz functions of the corresponding order).

Proposition 2.11. *If f, α and λ are as in Definition 2.10, we have for any $s \geq 0$ and $A \subset X$ that*

$$\mathcal{H}^{s/\alpha}(f(A)) \leq \lambda^{s/\alpha} \mathcal{H}^s(A).$$

Proof. Let $(C_i)_{i=1}^{\infty}$ be a δ-covering of A (with $\delta < c$, resp.). The Hölder continuity of f implies

$$|f(C_i)| \leq \lambda |C_i|^{\alpha} \leq \lambda \delta^{\alpha} =: \delta'.$$

Then $(f(C_i))_{i=1}^{\infty}$ forms a δ'-covering of $f(A)$ and

$$\sum_{i=1}^{\infty} |f(C_i)|^{s/\alpha} \le \lambda^{s/\alpha} \sum_{i=1}^{\infty} |C_i|^s .$$

Consequently, $\mathcal{H}_{\delta'}^{s/\alpha}(A) \le \lambda^{s/\alpha}\mathcal{H}_{\delta}^s(A)$. The limit as $\delta \to 0$ yields the assertion. □

The Hausdorff measures $\mathcal{H}^s$ of fractional dimensions s and their refinements $\mathcal{H}^h$ have been introduced for a quantitative description of geometric sets with highly irregular structure. In order to study properties of $\mathcal{H}^h$ on certain classes of fractal sets, *Vitali-type covering results* appear extremely useful. They will be treated in the next section.

2.2　Covering theorems

We start with a covering theorem for balls in a separable metric space. This case is well known from the literature. Its optimal version has been proved basically in [Mattila and Mauldin (1997)].

Theorem 2.12. *Let $\mathcal{B}$ be a family of closed balls in a separable metric space (X, d) with $\sup\{|B| : B \in \mathcal{B}\} < \infty$. Suppose that $A \subset X$ is such that each $x \in A$ is the centre of some $B \in \mathcal{B}$. Then for any $\varrho > 2$ there exists a (finite or infinite) sequence $(B(x_i, r_i))$ of (disjoint) balls from $\mathcal{B}$ such that $d(x_i, x_j) > r_i + r_j$, $i \ne j$, and*

$$A \subset \bigcup_i B(x_i, \varrho r_i) .$$

Proof. For each $x \in A$ choose some $B(x, r(x)) \in \mathcal{B}$. Set

$$M := \sup\{r(x) : x \in A\}.$$

Let $2/\varrho < \theta < 1$ and define

$$A_1 := \{x \in A : \theta M < r(x) \le M\}.$$

Choose a maximal subset (in the sense of Zorn's lemma for set inclusion) B_1 of A_1 such that $x, y \in B_1$, $x \ne y$, implies $x \notin B(y, \varrho r(y))$ or $y \notin B(x, \varrho r(x))$. Then we get $A_1 \subset \bigcup_{x \in B_1} B(x, \varrho r(x))$, since otherwise B_1 would not be maximal. Furthermore, for $x, y \in B_1$, $x \ne y$, by construction,

$$d(x, y) > \varrho \min\{r(x), r(y)\} > \varrho\,\theta\,M > 2M \ge r(x) + r(y)$$

and therefore the balls $B(x, r(x))$ and $B(y, r(y))$ are disjoint. Now let

$$A_2 := \left\{ x \in A \setminus \bigcup_{z \in B_1} B(z, \varrho r(z)) : \theta^2 M < r(x) \le \theta M \right\}.$$

Choose a maximal subset $B_2 \subset A_2$ such that for any $x, y \in B_2$ with $x \ne y$ it follows $x \notin B(y, \varrho r(y))$ or $y \notin B(x, \varrho r(x))$. Using the first step and similar arguments for the second step we conclude that

$$A_1 \cup A_2 \subset \bigcup_{x \in B_1 \cup B_2} B(x, \varrho r(x))$$

and the balls $B(x, r(x))$, $x \in B_1 \cup B_2$, satisfy the desired inequalities.

Iterating this process we obtain a family of balls $B(x, r(x))$, $x \in \bigcup_k B_k$, from $\mathcal{B}$ satisfying the desired inequalities such that the balls $B(x, \varrho r(x))$ cover A. By the separability of X this family is countable. $\qquad\Box$

We now turn to *fine covers* of the space.

Definition 2.13. A family $\mathcal{V}$ of subsets of X is called *Vitali class* for $A \subset X$ if for any $x \in A$ and $\delta > 0$ there exists some $V \in \mathcal{V}$ such that $x \in V$ and $0 < |V| < \delta$.

The next result, known as *Vitali covering theorem* has many applications in geometric measure theory and fractal geometry. Recall Definition (2.1) of the system $\mathbb{H}$ of gauge functions.

Theorem 2.14. *Let A be a subset of a metric space X and $\delta_0 > 0$. Suppose that $h \in \mathbb{H}$ satisfies the doubling condition $h(2u) \le Kh(u)$, $u > 0$, for some constant $K > 0$.*

(i) *If $\mathcal{V}$ is a Vitali class of closed sets for A, then there exists a sequence $(V_i)_i$ of disjoint sets from $\mathcal{V}$ such that either $\sum_{i=1}^{\infty} h(|V_i|) = \infty$, or*
$$\mathcal{H}_\delta^h\left(A \setminus \bigcup_i V_i\right) = 0 \text{ for all } \delta \le \delta_0 \text{ and hence, } \mathcal{H}^h\left(A \setminus \bigcup_i V_i\right) = 0.$$

(ii) *If $\mathcal{B}$ is a Vitali class of closed balls for A, then there exists a sequence $(B(x_i, r_i))$ of disjoint balls from $\mathcal{B}$ such that either $\sum_{i=1}^{\infty} h(|B(x_i, r_i)|) = \infty$, or for all $\delta < \delta_0$,*
$$\mathcal{S}_\delta^h\left(A \setminus \bigcup_i B(x_i, r_i)\right) = 0, \text{ and hence, } \mathcal{S}^h\left(A \setminus \bigcup_i B(x_i, r_i)\right) = 0.$$

(iii) *If $\mathcal{H}^h(A) < \infty$ (or $\mathcal{S}^h(A) < \infty$), then for any $\varepsilon > 0$ the sequence (V_i) in (i) (or $(B(x_i, r_i))$ in (ii)) may chosen so that*
$$\mathcal{H}^h(A) < \sum_i h(|V_i|) + \varepsilon \ \left(\text{or } \mathcal{S}^h(A) < \sum_i h(|B(x_i, r_i)|) + \varepsilon\right).$$

Proof. Without loss of generality we may assume $0 < |V| < \delta_0$, for any $V \in \mathcal{V}$. The sets V_i will be determined inductively:

Take an arbitrary $V_1 \in \mathcal{V}$ and suppose the $V_1, \ldots, V_k$ are chosen. Set

$$d_k := \sup \left\{ |W| : W \in \mathcal{V}, W \cap \bigcup_{i=1}^{k} V_i = \emptyset \right\}.$$

$d_k = 0$ means $A \subset \bigcup_{i=1}^{k} V_i$, since the last set is closed and $\mathcal{V}$ is a Vitali class. Therefore (i) is fulfilled for a finite sequence $V_1, \ldots, V_k$.

If $d_k > 0$ there exists some $V_{k+1} \in \mathcal{V}$ such that $V_{k+1} \cap \bigcup_{i=1}^{k} V_i = \emptyset$ and $|V_{k+1}| > d_k/2$. Thus, by induction, we obtain a sequence of sets $V_i \subset \mathcal{V}$ which is either finite or for any i, $|V_{i+1}| > d_i/2 > 0$.

If this sequence is infinite and $\sum_{i=1}^{\infty} h(|V_i|) < \infty$ we will show that $\mathcal{H}_\delta^h \big(A \setminus \bigcup_{i=1}^{\infty} V_i \big) = 0$, $\delta \le \delta_0$. For we first prove that for any $x_i \in V_i$, where $i \in \mathbb{N}$, and any $k \in \mathbb{N}$,

$$A \setminus \bigcup_{i=1}^{k} V_i \subset \bigcup_{i=k+1}^{\infty} B(x_i, 3|V_i|), \qquad (2.6)$$

which is of independent interest. If $x \in A \setminus \bigcup_{n=1}^{k} V_n$, then there exists some $V \in \mathcal{V}$ containing x such that $V \cap \bigcup_{n=1}^{k} V_n = \emptyset$ by closeness of $\bigcup_{n=1}^{k} V_n$ and the Vitali property of $\mathcal{V}$. Due to the convergence of the series $\sum_n h(|V_n|)$ we infer that $\lim_{n \to 0} |V_n| = 0$, since $h(u) > 0$ for $u > 0$. Thus, for some $m > k$, we get

$$|V| > 2|V_m| > d_{m-1} = \sup \left\{ |W| : W \cap \bigcup_{n=1}^{m-1} V_n = \emptyset \right\} > 0.$$

Hence, there exists some $i \in \{k+1, \ldots, m-1\}$ such that $V \cap V_i \neq \emptyset$ and $V \cap \bigcup_{n=1}^{i-1} V_n = \emptyset$. The definition of d_{i-1} yields $|V| \le d_{i-1} < 2|V_i|$ and therefore $V \subset B(x_i, 3|V_i|)$. Taking into regard that x is an arbitrary element of V we obtain (2.6).

Note that in case (ii), where the Vitali system consists of balls, the numbers d_k may be replaced by

$$d_k := \sup \left\{ r : B(x, r) \in \mathcal{B}, B(x, r) \cap \bigcup_{i=1}^{k} B(x_i, r_i) = \emptyset \right\},$$

and (2.6) takes the form

$$A \setminus \bigcup_{i=1}^{k} B(x_i, r_i) \subset \bigcup_{i=k+1}^{\infty} B(x_i, 3r_i)$$

(the arguments are completely analogous).

Tuning back to (i) and using (2.6) together with the doubling condition for h we estimate

$$\mathcal{H}^h_\delta\Big(A \setminus \bigcup_{i=1}^\infty V_i\Big) \le \mathcal{H}^h_\delta\Big(A \setminus \bigcup_{i=1}^k V_i\Big) \le \sum_{i=k+1}^\infty h(|B(x_i, 3|V_i|)|)$$

$$\le K^3 \sum_{i=k+1}^\infty h(|V_i|)$$

for all sufficiently large k such that $6|V_i| < \delta$ if $i \ge k+1$. Hence,

$$\mathcal{H}^h_\delta\Big(A \setminus \bigcup_{i=1}^\infty V_i\Big) \le K^3 \lim_{k \to \infty} \sum_{i=k+1}^\infty h(|V_i|) = 0,$$

which completes the proof of (i). The arguments for $\mathcal{S}^h_\delta$ are similar.

In order to show (iii) we choose δ_0 so small that

$$\mathcal{H}^h(A) < \mathcal{H}^h_{\delta_0}(A) + \varepsilon \le \mathcal{H}^h_{\delta_0}\Big(A \setminus \bigcup_{i=1}^\infty V_i\Big) + \mathcal{H}^h_{\delta_0}\Big(\bigcup_{i=1}^\infty V_i\Big) + \varepsilon$$

$$= \mathcal{H}^h_{\delta_0}\Big(\bigcup_{i=1}^\infty V_i\Big) + \varepsilon \le \sum_{i=1}^\infty h(|V_i|) + \varepsilon,$$

and similarly for $\mathcal{S}^h$. $\qquad\square$

As a first application of the Vitali theorem we get the following result for $X = \mathbb{R}^n$.

Theorem 2.15. *For any $\delta > 0$,*

$$2^n \omega_n^{-1} \mathcal{L}^n = \mathcal{H}^n = \mathcal{H}^n_\delta = \mathcal{S}^n_\delta = \mathcal{S}^n,$$

where $\omega_n := \Gamma(\frac{1}{2})^n / \Gamma(\frac{n}{2} + 1)$ is the volume of the unit ball in $\mathbb{R}^n$.

Proof. Fix $\delta > 0$. We first will show that

$$2^n \omega_n^{-1} \mathcal{L}^n(A) \le \mathcal{H}^n_\delta(A). \tag{2.7}$$

It suffices to consider the case $\mathcal{H}^n_\delta(A) < \infty$. For any $\varepsilon > 0$ choose a δ-covering $(C_i)_i$ of A such that

$$\sum_{i=1}^\infty |C_i|^n < \mathcal{H}^n_\delta(A) + \varepsilon.$$

Without loss of generality we may assume that the sets C_i are convex, since the diameter of C_i equals that of the convex hull of C_i. By the isodiametric inequality of convex geometry we have

$$\mathcal{L}^n(C_i) \le 2^{-n}\omega_n |C_i|^n \,,$$

where the right-hand side agrees with the volume of a ball of the same diameter $|C_i|$. (See, e.g., [Federer (1969)].) Thus,

$$\mathcal{L}^n(A) \le \sum_{i=1}^{\infty} \mathcal{L}^n(C_i) \le 2^{-n}\omega_n \sum_{i=1}^{\infty} |C_i|^n \le 2^{-n}\omega_n (\mathcal{H}^n_\delta(A) + \varepsilon) \,,$$

which yields (2.7).

On the other hand, there exists a sequence $(R_i)_{i=1}^{\infty}$ of open cuboids such that

$$\sum_{i=1}^{\infty} \mathcal{L}^n(R_i) < \mathcal{L}^n(A) + \varepsilon \,.$$

For any $i \in \mathbb{N}$ the closed balls inside R_i with diameters less than δ form a Vitali system for R_i. By Theorem 2.14 (ii) there exists a subsequence $(B_{ij})_j$ of disjoint balls with the property $\mathcal{S}^n_\delta(R_i \setminus \bigcup_j B_{ij}) = 0$, since

$$2^{-n}\omega_n \sum_j |B_{ij}|^n = \mathcal{L}^n\Big(\bigcup_j B_{ij}\Big) \le \mathcal{L}^n(R_i) < \infty \,.$$

Hence,

$$\begin{aligned}
\mathcal{S}^n_\delta(A) &\le \sum_{i=1}^{\infty} \mathcal{S}^n_\delta(R_i) \le \sum_{i=1}^{\infty} \mathcal{S}^n_\delta\Big(R_i \setminus \bigcup_j B_{ij}\Big) + \sum_{i=1}^{\infty} \mathcal{S}^n_\delta\Big(\bigcup_j B_{ij}\Big) \\
&= \sum_{i=1}^{\infty} \mathcal{S}^n_\delta\Big(\bigcup_j B_{ij}\Big) \le \sum_{i=1}^{\infty}\sum_j \mathcal{S}^n_\delta(B_{ij}) \le \sum_{i=1}^{\infty}\sum_j |B_{ij}|^n \\
&= 2^n\omega_n^{-1} \sum_{i=1}^{\infty}\sum_j \mathcal{L}^n(B_{ij}) = 2^n\omega_n^{-1} \sum_{i=1}^{\infty} \mathcal{L}^n\Big(\bigcup_j B_{ij}\Big) \\
&\le 2^n\omega_n^{-1} \sum_{i=1}^{\infty} \mathcal{L}^n(R_i) = 2^n\omega_n^{-1} \sum_{i=1}^{\infty} \mathcal{L}^n(R_i) \\
&< 2^n\omega_n^{-1}(\mathcal{L}^n(A) + \varepsilon) \,.
\end{aligned}$$

Since $\varepsilon > 0$ was arbitrary, the last chain of inequalities leads to

$$\mathcal{S}^n_\delta(A) \le 2^n\omega_n^{-1}\mathcal{L}^n(A) \,.$$

From this, (2.7) and $\mathcal{H}^n_\delta(A) \le \mathcal{S}^n_\delta(A)$ we conclude that

$$2^n\omega_n^{-1}\mathcal{L}^n(A) = \mathcal{H}^n_\delta(A) = \mathcal{S}^n_\delta(A)$$

for any $\delta > 0$, and hence also for the limit as $\delta \to 0$. $\qquad\square$

2.3 Packing measures in metric spaces

Packing measures in $\mathbb{R}^n$ have been introduced in 1982/85 by Tricot and Taylor as a certain counterpart to spherical and Hausdorff measures (see [Taylor and Tricot (1985)]). Instead of optimal coverings, optimal *ball packings* are used as a main tool. Here (X, d) is an arbitrary metric space.

The system of *gauge functions* $\mathbb{H}_-$ is defined now as $\mathbb{H}$ in (2.1), where right continuity is replaced by *left continuity:*

$$\mathbb{H}_- := \{h : [0, \infty] \to [0, \infty] : h(u) > 0, \text{ for } u > 0, \qquad (2.8)$$

$$h \text{ is left continuous and non-decreasing}\} . \qquad (2.9)$$

(Note that these continuities lead to Borel regularity of the Hausdorff and packing measures, resp.)

Definition 2.16. A *metric δ-ball packing* of $A \subset X$ is a sequence $(B(x_i, r_i))_{i \leq \infty}$ of closed balls such that:

(i) $x_i \in A$,
(ii) $d(x_i, x_j) > r_i + r_j$, for $i \neq j$,
(iii) $2r_i < \delta$.

Remark 2.17. Condition (ii) implies that the balls are disjoint. In $\mathbb{R}^n$ the converse is also true. However in so-called *ultrametric spaces*, where $d(x, z) \leq \max\{d(x, y), d(y, z)\}$, it can happen that $d(x_1, x_2) < r_1 + r_2$ and $B(x_1, r_1) \cap B(x_2, r_2) = \emptyset$. (From the above ultrametric triangle inequality it follows that two balls $B(x, r)$ and $B(y, r)$ are disjoint if and only if $d(x, y) > r$.)

If (ii) is replaced by the disjointness condition of the balls we will speak of (ordinary) *δ-ball packings*.

An advantage of Definition 2.16 (due to [Haase (1990)]) is, e.g., the better behavior under Lipschitz mappings, in particular, a natural non-expanding property of packing measures. Both types of packing are said to be *radii-based*. A *diameter-based δ-ball packing* is a sequence of disjoint balls with centers in A whose diameters are less than δ. This last concept involves some difficulties concerning the regularity of the associated packing measures (cf. [Joyce (1999)]).

In $\mathbb{R}^n$, however, $|B(x, r)| = 2r$ and hence, the packing systems agree.

For any gauge function $h \in \mathbb{H}_-$ we now define (*δ-approximate*) *packing premeasures* on X.

Definition 2.18. For $\delta > 0$ and $A \subset X$ we set (for $\sup(\emptyset) := 0$)

$$\mathcal{P}_\delta^h(A) := \sup\left\{ \sum_i h(2r_i) : (B(x_i, r_i))_i \text{ is a metric } \delta\text{-ball packing for } A \right\},$$

$$\widetilde{\mathcal{P}}_\delta^h(A) := \sup\left\{ \sum_{i=1}^\infty h(2r_i) : (B(x_i, r_i))_i \text{ is a } \delta\text{-ball packing for } A \right\},$$

$$\mathcal{P}_0^h(A) := \lim_{\delta \to 0} \mathcal{P}_\delta^h(A) = \inf_{\delta > 0} \mathcal{P}_\delta^h(A),$$

$$\widetilde{\mathcal{P}}_0^h(A) := \lim_{\delta \to 0} \widetilde{\mathcal{P}}_\delta^h(A) = \inf_{\delta > 0} \widetilde{\mathcal{P}}_\delta^h(A).$$

Observe that $\mathcal{P}_\delta^h(A)$ and $\widetilde{\mathcal{P}}_\delta^h(A)$ are non-increasing for decreasing δ and

$$\mathcal{P}_\delta^h(A) \le \widetilde{\mathcal{P}}_\delta^h(A).$$

As usual for $h(u) = u^s$ we write $\mathcal{P}_\delta^s(A)$, $\mathcal{P}_0^s(A)$, $\widetilde{\mathcal{P}}_\delta^s(A)$ and $\widetilde{\mathcal{P}}_0^s(A)$, respectively. The packing premeasures $\mathcal{P}_0^h$ and $\widetilde{\mathcal{P}}_0^h$ possess the following properties.

Exercise 2.1.

$$\mathcal{P}_0^h(A) \le \mathcal{P}_0^h(B) + \mathcal{P}_0^h(C), \quad \text{if } A \subset B \cup C$$

(analogously for $\widetilde{\mathcal{P}}_0^h$).

Exercise 2.2.

$$\mathcal{P}_0^h(A \cup B) = \mathcal{P}_0^h(A) + \mathcal{P}_0^h(B)$$

for positive separated sets A and B (analogously for $\widetilde{\mathcal{P}}_0^h$).

For both these relationships the left continuity of h is not needed, whereas it is essentially exploited in the next result.

Proposition 2.19. *For any $h \in \mathbb{H}_-$ and $A \subset X$ we have*

$$\mathcal{P}_0^h(A) = \mathcal{P}_0^h(\bar{A}),$$

(analogously for $\widetilde{\mathcal{P}}_0^h$).

Proof. Obviously, $\mathcal{P}_0^h(A) \le \mathcal{P}_0^h(\bar{A})$, since any metric δ-ball packing for A is also one for the closure $\bar{A}$.

Let now $(B(x_i, r_i))_i$ be a metric δ-ball packing for $\bar{A}$. For any $0 < \delta_i < \delta$ there exists some $y_i \in A$ such that $d(x_i, y_i) < \delta_i$. Choose $\delta_i < r_i$ and let $\varrho_i := r_i - \delta_i$. The triangle inequality applied twice yields the following:

$$d(y_i, y_j) > d(x_i, x_j) - \delta_i - \delta_j > r_i + r_j - \delta_i - \delta_j = \varrho_i + \varrho_j.$$

Hence, $(B(x_i, \varrho_i))_i$ is a metric δ-ball packing for A and

$$\mathcal{P}_\delta^h(A) \geq \sum_i h(2\varrho_i) = \sum_i h(2r_i - 2\delta_i).$$

By left continuity of h, for any $\varepsilon > 0$ we may choose δ_i so small that $h(2r_i - 2\delta_i) \geq h(2r_i) - \varepsilon/2^i$. Then we obtain $\sum_i h(2r_i - 2\delta_i) \geq \sum_i h(2r_i) - \varepsilon$. This implies

$$\sum_i h(2r_i) \leq \mathcal{P}_\delta^h(A) + \varepsilon$$

for any metric δ-ball packing $(B(x_i, r_i))_i$ of $\bar{A}$ and hence,

$$\mathcal{P}_\delta^h(\bar{A}) \leq \mathcal{P}_\delta^h(A) + \varepsilon.$$

Letting $\varepsilon \to 0$ and then $\delta \to 0$ we obtain $\mathcal{P}_0^h(\bar{A}) \leq \mathcal{P}_0^h(A)$.
    The proof for $\widetilde{\mathcal{P}}_0^h(A)$ is similar. $\qquad\square$

We now turn to the *non-expanding property* of $\mathcal{P}_0^h$ mentioned above. This property is not shared, in general, by $\widetilde{\mathcal{P}}_0^h$.

Proposition 2.20. *If* $f : (X, d) \to (X', d')$ *satisfies* $d'(f(x), f(y)) \leq d(x, y)$, *for* $d(x, y) < c$, *and some* $c > 0$, *then we have* $\mathcal{P}_0^h(f(A)) \leq \mathcal{P}_0^h(A)$.

Proof. It suffices to prove that for $\delta < c$ any metric δ-ball packing $(B(y_i, r_i))_i$ for $f(A)$ generates one for A with the same radii. For $y_i \in f(A)$ choose some $x_i \in A$ with $f(x_i) = y_i$. Then the sequence of balls $(B(x_i, r_i))_i$ forms a δ-ball packing for A, since

$$r_i + r_j < d'(y_i, y_j) = d'(f(x_i), f(x_j)) \leq < d(x_i, x_j).$$

$\qquad\square$

Exercise 2.3. If f is locally uniformly Lipschitz with constant $\lambda > 0$, i.e., $d(x, y) < c$ implies $d'(f(x), f(y)) \leq \lambda d(x, y)$, then we have

$$\mathcal{P}_0^s(f(A)) \leq \lambda^s \mathcal{P}_0^s(A), \quad s \geq 0.$$

Note that 2.1, 2.2 and 2.20 also hold for the Hausdorff measure $\mathcal{H}^h$, but $\mathcal{H}^h(A) = \mathcal{H}^h(\bar{A})$ is not true, in general. Nevertheless, packing premeasures behave worse than Hausdorff measures. The following example shows that *they are not necessarily σ-subadditive*.

Example 2.21. Let A be a countable dense subset of $\mathbb{R}^n$, i.e., $\bar{A} = \mathbb{R}^n$, and $0 < s \leq n$. In view of 2.19 we get

$$\mathcal{P}_0^s(A) = \mathcal{P}_0^s(\mathbb{R}^n) = \infty.$$

On the other hand , using $\mathcal{P}_0^s(\{x\}) = 0$, for any $x \in \mathbb{R}^n$ (which easily follows from the definition of $\mathcal{P}_0^h$) we infer

$$\sum_{a \in A} \mathcal{P}_0^s(\{a\}) = 0 \,.$$

Consequently,

$$\mathcal{P}_0^s(A) > \sum_{a \in A} \mathcal{P}_0^s(\{a\}) \,.$$

This disadvantage of the set function $\mathcal{P}_0^h$ has led to the following natural improvement of the construction in the sense of 1.12:

Definition 2.22.

$$\mathcal{P}^h(A) := \inf \left\{ \sum_{i=1}^{\infty} \mathcal{P}_0^h(A_i) : A = \bigcup_{i=1}^{\infty} A_i \right\}$$

is called *packing measure of $A \subset X$ with respect to the gauge function h,* briefly *packing h-measure.* $\widetilde{\mathcal{P}}^h$ is defined similarly.

For $h(u) = u^s$ we write $\mathcal{P}^s$ and $\widetilde{\mathcal{P}}^s$ and we call them *s-dimensional packing measures.* As a corollary of 1.12 and 2.2 we obtain:

Proposition 2.23. $\mathcal{P}^h$ *and* $\widetilde{\mathcal{P}}^h$ *are metric outer measures and therefore Borel measures.*

Note that $\mathcal{P}^h = \widetilde{\mathcal{P}}^h$ in $\mathbb{R}^n$ and, in general, one can estimate as follows.

Proposition 2.24.

$$\mathcal{P}^h(A) \le \widetilde{\mathcal{P}}^h(A) \,.$$

(Recall that this inequality already holds for the corresponding approximate packing premeasures $\mathcal{P}_\delta^h$ and $\widetilde{\mathcal{P}}_\delta^h$). Conversely, we get the following:

Proposition 2.25. *If h satisfies the doubling condition $h(2u) \le Kh(u)$ for $u > 0$, then we have*

$$\widetilde{\mathcal{P}}^h(A) \le K\mathcal{P}^h(A) \,.$$

Proof. Let $\{B(x_i, r_i)\}$ be a δ-ball packing for A. Using $d(x_i, x_j) \ge \max\{r_i, r_j\} \ge r_i/2 + r_j/2$ we obtain a metric $(\delta/2)$-ball packing $(B(x_i, r_i/2))_i$ in the sense of Definition 2.16. Since $h(2r_i) \le Kh(r_i)$, we infer for any $\delta > 0$,

$$\widetilde{\mathcal{P}}_\delta^h(A) \le K\mathcal{P}_{\delta/2}^h(A) \,,$$

hence $\widetilde{\mathcal{P}}_0^h(A) \le K\mathcal{P}_0^h(A)$ and thus, by 2.22, the assertion. $\square$

Furthermore, 2.3 implies the *Lipschitz property:*

Proposition 2.26. *If $f : (X, d) \to (X', d')$ is locally uniformly Lipschitz with constant $\lambda > 0$, then we have*

$$\mathcal{P}^s(f(A)) \leq \lambda^s \mathcal{P}^s(A).$$

Whereas in Propositions 2.23–2.25 the left continuity assumed for h is not really needed, it plays an essential role for proving that in 2.22 the *covering sets A_i* may be chosen *closed.*

Proposition 2.27. *If $h \in \mathbb{H}_-$, then*

$$\mathcal{P}^h(A) = \inf \left\{ \sum_{i=1}^{\infty} \mathcal{P}_0^h(F_i) : F_i \text{ closed}, A \subset \bigcup_{i=1}^{\infty} F_i \right\},$$

and the same holds for $\widetilde{\mathcal{P}}^h$.

Proof. By definition of $\mathcal{P}^h(A)$, for any $\varepsilon > 0$ there exists a sequence $(A)_{i=1}^{\infty}$ such that $A = \bigcup_{i=1}^{\infty} A_i$ and

$$\mathcal{P}^h(A) \geq \sum_{i=1}^{\infty} \sum \mathcal{P}_0^h(A_i) - \varepsilon = \sum_{i=1}^{\infty} \mathcal{P}_0^h(\bar{A}_i) - \varepsilon$$

according to Proposition 2.19. Taking into regard that $\sum_{i=1}^{\infty} \mathcal{P}_0^h(\bar{A}_i)$ is not less than the right hand side of the asserted equality we infer that the latter does not exceed $\mathcal{P}^h(A)$. The opposite inequality is obvious. The same arguments hold also for $\widetilde{\mathcal{P}}^h$. □

The last result is the key for proving the *Borel regularity* of packing measures.

Corollary 2.28. *For any gauge function $h \in \mathbb{H}_-$ the packing measures $\mathcal{P}^h$ and $\widetilde{\mathcal{P}}^h$ are Borel regular.*

Proof. The arguments for $\mathcal{P}^h$ and $\widetilde{\mathcal{P}}^h$ are the same. By 2.27 for any $n \in \mathbb{N}$ there exists a covering of A by closed sets $(F_i^n)_{i=1}^{\infty}$ such that

$$\sum_{i=1}^{\infty} \mathcal{P}_0^h(F_i^n) < \mathcal{P}^h(A) + \frac{1}{n}.$$

For the Borel set $B := \bigcap_{n=1}^{\infty} \bigcup_{i=1}^{\infty} F_i^n \supset A$ and any n one obtains

$$\mathcal{P}^h(A) \leq \mathcal{P}^h(B) \leq \mathcal{P}^h\left(\bigcup_{i=1}^{\infty} F_i^n\right) \leq \sum_{i=1}^{\infty} \mathcal{P}_0^h(F_i^n) < \mathcal{P}^h(A) + \frac{1}{n}.$$

Letting $n \to \infty$ we conclude $\mathcal{P}^h(A) = \mathcal{P}^h(B)$. □

2.4   An optimal Vitali theorem and comparison of Hausdorff and packing measures

As we will see later Hausdorff and packing measures may be different on sets of highly irregular structure. In this section we will derive an estimate. As a useful tool a Vitali-type theorem for ball packings will be proved.

Recall that in Theorem 2.12 and in the proof of Theorem 2.14 we have obtained ball packings $(B(x_i, r_i))_i$ for a given set such that under certain conditions the balls with the same centers and radii $3r_i$ cover the whole set and have the fine covering property (2.7), respectively. In the first case small balls do not play a role, but the space is assumed to be separable. In the second case we have used estimates valid for general Vitali systems, since the constant 3 is not essential for those purposes. We now will improve this in the following sense.

Definition 2.29. A *centered Vitali system $\mathcal{B}$ of closed balls* for a nonempty set A in X means that any $x \in A$ is the center of balls from $\mathcal{B}$ with arbitrarily small radii.

Theorem 2.30. *Let X be a metric space, $\mathcal{B}$ be a centered Vitali system of closed balls for A and let $\varrho < 1/2$. Then there exists a countable subsystem of balls $(B(x_i, r_i))_i$ with*

$$d(x_i, x_j) > \varrho(r_i + r_j), \ i \neq j,$$

such that infinitely many r_i are greater than some positive constant or

$$A \setminus \bigcup_{i=1}^{k} B(x_i, \varrho r_i) \subset \bigcup_{i>k} B(x_i, r_i)$$

for any $k \geq 0$. In the first case we have $\sum_{i=1}^{\infty} h(2\varrho r_i) = \infty$. Otherwise

$$\mathcal{H}_{\delta}^{h}\left(A \setminus \bigcup_{i} B(x_i, \varrho r_i)\right) = 0$$

for any $\delta > 0$, provided h fulfills the doubling condition.

(For ordinary ball packings this result may be found in [Cutler (1995)].)

Proof. We determine the balls inductively.

Take $x_1 \in A$, $B(x_1, r_1) \in \mathcal{B}$. Suppose that $x_1, \ldots, x_k$ and $r_1, \ldots, r_k$ are chosen. Set

$$d_k := \sup\{r : \exists\, x \in A \text{ with } d(x, x_i) > \varrho(r + r_i), i = 1, \ldots, k, B(x, r) \in \mathcal{B}\}.$$

If $d_k = 0$, then there is no $x \in A$ with $d(x, x_i) > \varrho r_i$ for all i. In this case $A \subset \bigcup_{i=1}^{k} B(x_i, \varrho r_i)$ and we can stop (having the assertion for finitely many balls).

In case $d_k > 0$ choose $r_{k+1} > \theta d_k$, where $\theta := \varrho/(1 - \varrho) < 1$ and $x_{k+1} \in A$ such that $B(x_{k+1}, r_{k+1}) \in \mathcal{B}$ and $d(x_{k+1}, x_i) > \varrho(r_{k+1} + r_i)$, $i = 1, \ldots, k$.

If $d_k > 0$ for any integer $k \in \mathbb{N}$, we obtain an infinite ball packing $\{B(x_i, \varrho r_i)\}_{i=1}^{\infty}$ for A.

It remains to consider the case $\lim_{n \to \infty} r_n = 0$. For any $x \in A \setminus \bigcup_{i=1}^{k} B(x_i, \varrho r_i)$, there is some $r > 0$ satisfying $d(x, x_i) > \varrho(r + r_i)$, $i = 1, \ldots, k$, and $B(x, r) \in \mathcal{B}$. ($d(x, x_i) > r_i + \varepsilon_i$ for some $\varepsilon_i > 0$, $i = 1, \ldots, k$, therefore we can choose some $r \leq \min_i \varepsilon_i$.) For sufficiently large $n > k$ we have

$$r > \theta^{-1} r_n > d_{n-1}.$$

By the definition of d_{n-1} there exists the smallest $i \in \{k + 1, \ldots, n - 1\}$ such that

$$d(x, x_i) \leq \varrho(r + r_i)$$

(otherwise $r \leq d_{n-1}$). Minimality of i implies $d(x, x_j) > \varrho(r + r_j)$, $j \leq i - 1$, in particular, r is a candidate in the definition of d_{i-1}. Consequently,

$$r \leq d_{i-1} < \theta^{-1} r_i$$

and hence,

$$d(x, x_i) < \varrho(\theta^{-1} + 1)r_i = r_i$$

i.e., $x \in B(x_i, r_i)$. Thus, we have shown that

$$A \setminus \bigcup_{i=1}^{k} B(x_i, \varrho r_i) \subset \bigcup_{k+1}^{\infty} B(x_i, r_i).$$

The rest is the same as in the proof of Theorem 2.14. □

This theorem enables us *to compare Hausdorff and packing measures in metric spaces*. Here the continuity properties of the gauge functions are not needed:

Corollary 2.31. *If the gauge function h satisfies the doubling condition, then we have*

$$\mathcal{H}^h(A) \leq \mathcal{P}^h(A), \quad A \subset X.$$

Proof. It suffices to show $\mathcal{H}^h(A) \leq \mathcal{P}_0^h(A)$, since this implies

$$\mathcal{H}^h(A) \leq \sum_{i=1}^{\infty} \mathcal{H}^h(A_i) \leq \sum_{i=1}^{\infty} \mathcal{P}_0^h(A_i), \quad A \subset \bigcup_{i=1}^{\infty} A_i,$$

hence, $\mathcal{H}^h(A) \leq \mathcal{P}^h(A)$. We may assume $\mathcal{P}_0^h(A) < \infty$. Then there exists some δ_0 such that $\mathcal{P}_\delta^h(A) < \infty$ for all $\delta < \delta_0$. For an arbitrary metric δ-ball packing $(B(x_i, \varrho_i))_i$ for A we have

$$\sum_i h(2\varrho_i) \leq \mathcal{P}_\delta^h(A) < \infty. \tag{2.10}$$

Therefore Theorem 2.30 provides for $\varrho < 1/2$ some metric δ-ball packing $(B(x_i, \varrho r_i))_i$ for A satisfying

$$\mathcal{H}_\delta^h\big(A \setminus \bigcup_i B(x_i, \varrho r_i)\big) = 0.$$

This and (2.10) imply

$$\mathcal{H}_\delta^h(A) \leq \mathcal{H}_\delta^h\big(A \setminus \bigcup_i B(x_i, \varrho r_i)\big) + \mathcal{H}_\delta^h\big(\bigcup_i B(x_i, \varrho r_i)\big)$$

$$= \mathcal{H}_\delta^h\big(\bigcup_i B(x_i, \varrho r_i)\big) \leq \sum_i h(2\varrho r_i) \leq \mathcal{P}_\delta^h(A).$$

Letting $\delta \to 0$ the assertion follows. $\qquad\qquad\qquad\qquad\qquad\square$

2.5 Hausdorff and packing dimensions

We have already used the notions of s-dimensional Hausdorff and packing measures. In this section we will present some basic ideas of measure theoretic dimension theory, which is an important part of modern fractal geometry.

Familiar dimension concepts are those of linear algebra (dimension of vector spaces), topology (topological dimension) and differential geometry (dimension of manifolds). The latter may also be interpreted in a measure geometric context: If M_m is a smooth m-dimensional submanifold of $\mathbb{R}^n$, $m < n$, then the Lebesgue measure $\mathcal{L}^n$ restricted to M_m vanishes. The usual m-dimensional surface area measure, which agrees with $\mathcal{H}^m \llcorner M_m$, is positive and locally finite. The above disadvantage of $\mathcal{L}^n$ carries over to all $p \in \{1, \ldots, m-1, m+1, \ldots, n\}$. It is not difficult to see that

$$\mathcal{H}^p \llcorner M_m = \begin{cases} +\infty, & \text{if } p < m \\ 0, & \text{if } p > m, \end{cases}$$

where $+\infty$ stands for the values on relatively open sets. This observation has a "fractional" generalization which will be demonstrated for Hausdorff measures as well as for packing measures (extensions to other suitable measures are straightforward).

Turning back to general metric space (X, d) we first prove a *jump property* of Hausdorff and packing measures for the gauge functions $h(u) = u^s$.

Proposition 2.32. *Let $A \subset X$ and $s \geq 0$. μ^s stands for $\mathcal{H}^s$, $\mathcal{P}^s$ or $\widetilde{\mathcal{P}}^s$. Then $\mu^s(A) < \infty$ implies $\mu^t(A) = 0$ for any $t > s$.*

Proof. For any $\delta > 0$ we can choose a δ-covering $(C_i)_{i=1}^{\infty}$ of A such that

$$\sum_{i=1}^{\infty} |C_i|^s < \mathcal{H}^s(A) + 1.$$

Estimating

$$\mathcal{H}_\delta^t(A) \leq \sum_{i=1}^{\infty} |C_i|^t = \sum_{i=1}^{\infty} |C_i|^{t-s} |C_i|^s \leq \delta^{t-s} \sum_{i=1}^{\infty} |C_i|^s$$
$$< \delta^{t-s} (\mathcal{H}^s(A) + 1)$$

and letting $\delta \to 0$ we obtain the assertion for $\mathcal{H}^s$.

The arguments for $\widetilde{\mathcal{P}}_0$ (resp. $\mathcal{P}_0$) are similar, using that $\sum_i |2r_i|^s \leq \widetilde{\mathcal{P}}_\delta^s(B)$ (or $\mathcal{P}_\delta^s(B)$) for any (metric) δ-ball packing $(B(x_i, r_i))_i$ of $B \subset X$. Hence, $\widetilde{\mathcal{P}}_0^t(B) = 0$ if $B \subset A$ and $\widetilde{\mathcal{P}}^s(B) < \infty$. Since $\mathcal{P}^s(A) < \infty$, we obtain from the definition of $\widetilde{\mathcal{P}}^t$ that

$$\widetilde{\mathcal{P}}^t(A) \leq \inf \left\{ \sum_{i=1}^{\infty} \widetilde{\mathcal{P}}_0^t(A_i) : A = \bigcup_{i=1}^{\infty} A_i, \ \widetilde{\mathcal{P}}_0^s(A_i) < \infty \right\} = 0$$

(take $\mathcal{P}_0^t$ and $\mathcal{P}^t$, resp.). $\qquad\square$

According to this property there is at most one point $s = D$ where the values of $\mu^s(A)$ for increasing s jump from $+\infty$ to 0. By the equivalence of $\mathcal{P}$ and $\widetilde{\mathcal{P}}$ (cf. 2.24 and 2.25) the jump point D is the same for $\mathcal{P}^s(A)$ and $\widetilde{\mathcal{P}}^s(A)$.

This point D will be called the corresponding dimension of the set A (as before, by convention, $\sup \emptyset = 0$ and $\inf \emptyset = +\infty$):

Definition 2.33. *Let $A \subset X$. Then*

$$\dim_H A := \sup\{s \geq 0 : \mathcal{H}^s(A) = \infty\} = \inf\{s \geq 0 : \mathcal{H}^s(A) = 0\}$$

is the Hausdorff dimension of A and

$$\dim_P A := \sup\{s \geq 0 : \mathcal{P}^s(A) = \infty\} = \inf\{s \geq 0 : \mathcal{P}^s(A) = 0\}$$

is its packing dimension.

An immediate consequence of Theorem 2.30 is the following:

Proposition 2.34.

$$\dim_H A \le \dim_P A, \, A \subset X.$$

Remarks 2.1. Below we will see examples where the inequality is strict. These are sets whose "fractal" behavior is rather irregular. For most of the sets which been studied in fractal geometry the topological dimension is less than the Hausdorff dimension. $\dim_H A = D$ or $\dim_P A = D$ does not imply anything on the values $\mathcal{H}^D(A)$ or $\mathcal{P}^D(A)$, respectively. Both measures and dimensions play an important role for invariant sets in the theory of dynamical systems.

Hausdorff measures and dimension have better Euclidean projection and section properties than the packing variants.

Concerning square products of sets packing measures and dimension are essential: For $X = \mathbb{R}^n$ it follows from Tricot [1982] that

$$\dim_H A + \dim_H B \le \dim_H(A \times B) \le \dim_P A + \dim_H B$$
$$\le \dim_P(A \times B) \le \dim_P A + \dim_P B.$$

For the special case $\dim_H A = \dim_P A$ this implies

$$\dim_H(A \times B) = \dim_H A + \dim_H B$$
$$\dim_P(A \times B) = \dim_P A + \dim_P B.$$

Moreover, in [Bishop and Peres (1996)] it is proved that

$$\dim_P A = \sup\{\dim_H(A \times B) - \dim_H B : B \text{ compact}\}$$

for any Borel (and more generally, Suslin) set A in $\mathbb{R}^n$.

We will work out some properties of Hausdorff and packing dimensions and apply them to certain classes of fractal sets. First note that for the dimension notions the measures are not really needed. For the Hausdorff dimension this is easy to see:

Exercise 2.4.

$$\dim_H A = \inf\{s \ge 0 : \text{ for any } \varepsilon > 0 \text{ there is a ball covering}$$
$$(B_i)_i \text{ of } A \text{ such that } \sum_{i=1}^{\infty} |B_i|^s < \varepsilon\}.$$

A simple characterization of packing dimension without packing measures will be given in Theorem 5.4.2.

Both Hausdorff and packing dimension are *monotone* and *σ-stable*:

Proposition 2.35.

    (i) $\dim A \leq \dim B$, *if* $A \subset B$,

    (ii) $\dim \left(\bigcup_{i=1}^{\infty} A_i \right) = \sup_i \dim A_i$.

(Roughly speaking (ii) says that the dimension of the set agrees with that of its "thickest" part.)

Proof. (i) follows from monotonicity of the outer measure $\mathcal{H}^s$ and $\mathcal{P}^s$, respectively. It implies

$$\sup_i \dim A_i \leq \dim \left(\bigcup_i A_i \right).$$

For the opposite inequality assume that $\sup_i \dim A_i < \infty$. If $\dim A_i < s$ for any i, then $\mu^s(A_i) = 0$ for any i, and hence $\mu^s \left(\bigcup_{i=1}^{\infty} A_i \right) = 0$, where μ^s stands for $\mathcal{H}^s$ or $\mathcal{P}^s$, respectively. This yields

$$\dim \left(\bigcup_{i=1}^{\infty} A_i \right) \leq s.$$

Putting $s := \sup_i \dim A_i + \varepsilon > \dim A_j$, $j \in \mathbb{N}$, we obtain

$$\dim \left(\bigcup_{i=1}^{\infty} A_i \right) \leq \sup_i \dim A_i + \varepsilon$$

and

$$\dim \left(\bigcup_{i=1}^{\infty} A_i \right) \leq \sup_i \dim A_i$$

letting $\varepsilon \to 0$. Thus, (ii) is proved. $\qquad\square$

The following *mapping properties* are well known:

Proposition 2.36. *Suppose* $f : (X, d) \to (X', d')$ *satisfies the Hölder condition* $d'(f(x), f(y)) \leq \lambda d(x, y)^{\alpha}$ *if* $d(x, y) < c$ *for some* $\alpha, c, \lambda > 0$. *Then we have for any* $A \subset X$,

    (i) $\dim_H f(A) \leq \alpha^{-1} \dim_H A$,

    (ii) $\dim_P f(A) \leq \dim_P A$ *for* $\alpha = 1$,

    (iii) $\dim_H f(A) = \dim_H A$ *if* f *is bijective and locally uniformly bi-Lipschitz.*

Proof. Choosing $s := \dim_H A + \varepsilon$ and applying 2.11 we obtain that $0 = \mathcal{H}^s(A) \geq \lambda^{-s/\alpha} \mathcal{H}^{s/\alpha}(f(A))$, i.e, $\mathcal{H}^{s/\alpha}(f(A)) = 0$, hence,

$$\dim_H f(A) \leq \alpha^{-1} s = \alpha^{-1} \dim_H A + \alpha^{-1} \varepsilon$$

for any $\varepsilon > 0$. This yields (i).

Exercise 2.3 implies $\mathcal{P}^s(f(A)) \leq \mathcal{P}^s(A)$ for $\alpha = 1$. From this (ii) follows similarly as above. (iii) is a consequence of (i). $\qquad\square$

Before studying methods of calculating or estimating dimensions we will demonstrate some basic ideas of determining Hausdorff dimension on the most frequently considered example of *Cantor sets*. This is the mathematically simplest variant of so called self-similar sets. On the other hand a non-linear version of Cantor sets plays an important role in physics when considering local intersections of strange attractors in dynamical systems with certain hyperplanes.

Example 2.37. Let $0 < a < 1/2$ and $S_1(x) := ax$, $S_2(x) := ax + 1 - a$, $x \in I := [0, 1]$, be the similarities mapping the interval I onto the contracted intervals $[0, a]$ and $[1 - a, 1]$. Define $C_0 = I$, $C_1 := S_1(C_0) \cup S_2(C_0)$, and inductively

$$C_{k+1} := S_1(C_k) \cup S_2(C_k).$$

In construction step k we get a finite union C_k of 2^k disjoint intervals of length a^k arising from the preceding set when deleting from the middle of each interval of C_{k-1} an open interval of length $(1 - 2a)a^{k-1}$. Then the limit set

$$C(a) := \bigcap_{k=1}^{\infty} C_k$$

is an uncountable compact set without interior points and with zero Lebesgue measure. The classical case with $a = 1/3$ leads to the so-called *middle-third Cantor set*. There is a simple intuitive argument for calculating the Hausdorff dimension of $C(a)$. For any $s \geq 0$ the invariance of $\mathcal{H}^s$ under translations and its scaling property (cf. 2.7) yield, together with the *self-similarity*

$$C = S_1(C) \cup S_2(C)$$

of the Cantor set $C = C(a)$, the following:

$$\mathcal{H}^s(C) = \mathcal{H}^s(S_1(C) \cup S_2(C)) = \mathcal{H}^s(S_1(C)) + \mathcal{H}^s(S_2(C))$$
$$= a^s \mathcal{H}^s(C) + a^s \mathcal{H}^s(C) = 2a^s \mathcal{H}^s(C).$$

If $D := \dim_H C$ and if we can show that $0 < \mathcal{H}^D(C) < \infty$, we consider the case $s = D$ and divide the above equations by $\mathcal{H}^s(C)$. Then we obtain $1 = 2a^D$, i.e.,

$$\dim_H C(a) = D = \frac{\log 2}{|\log a|}.$$

The upper estimate $\mathcal{H}^D(C) < \infty$ can easily be proved by choosing optimal coverings of C adapted to its construction (this idea extends to more general sets in dynamical systems): For any $k \in \mathbb{N}$ we take the 2^k closed intervals I_k^j of length a^k constituting C_k and obtain for D as above $\mathcal{H}_{a^k}^D(C) \le 2^k(a^k)^D = 1$, hence, $\mathcal{H}^D(C) \le 1$, letting $k \to \infty$.

Lower estimates of Hausdorff measures are, in general, more difficult. For Cantor sets, however, direct arguments show that $\mathcal{H}^D(C) \ge 1/4$: It suffices to prove that $\sum_{i=1}^\infty |G_i|^D \ge 1/4$ for any open covering $(G_i)_{i=1}\infty$ of C. Since C is compact we may choose a finite subcovering, say $G_1, \ldots, G_n$. For fixed $i \in \{1, \ldots, n\}$ define $l \in \mathbb{N}$ by

$$a^{l+1} \le |G_i| < a^l.$$

Then the number of intervals I_k^j of C_k which intersect G_i does not exceed $2^{k+1}/2^l$. This implies

$$\sum_{I_k^j \cap G_i \ne \emptyset} |I_k^j|^D \le 2^{k+1-l} a^{kD} = 2^{1-l} = 4a^{(l+1)D} \le 4|G_i|^D.$$

Summing up over i we infer

$$4 \sum_{i=1}^n |G_i|^D \ge \sum_{i=1}^n \sum_{I_k^j \cap G_i \ne \emptyset} |I_k^j|^D \ge \sum_{j=1}^{2^k} |I_k^j|^D = 2^k a^{kD} = 1.$$

Remark 2.38. By an intricate refinement of the last arguments one can show that $\mathcal{H}^D(C) \ge 1$ (cf. [Dvoretzky (1948)]). Since the opposite inequality has been proved above we get that $\mathcal{H}^D(C) = 1$.

2.6 Tightness of Hausdorff and packing measures

The approximation property of a finite Borel measure μ by means of compact subsets given in Theorem 1.25 is sometimes called tightness of μ.

Definition 2.39. A Borel measure μ in a metric space is said to be *tight* if for any Borel set A,

$$\mu(A) = \sup\{\mu(K) : K \text{ compact}, K \subset A, \mu(K) < \infty\}.$$

Theorem 2.40. *Let (X, d) be a complete separable metric space. Then the Hausdorff measure $\mathcal{H}^h$, $h \in \mathbb{H}$, and the packing measure $\mathcal{P}^h$, $h \in \mathbb{H}_- \cap \mathbb{H}$, are tight, provided h satisfies the doubling condition.*

Remarks 2.2. (on the historical development of the proof)

  (i) The case of Borel sets A with finite measure is already covered by Theorem 1.25.

(ii) For Euclidean spaces X Besicovitch proved in 1952 that any compact set A with $\mathcal{H}^s(A) > 0$ contains a compact subset K such that $0 < \mathcal{H}^s(K) < \infty$. In the same year Davies extended this to Suslin sets A, which are more general than Borel sets.

(iii) It took more than 40 years before [Howroyd (1995)] proved the Davies result for general complete separable metric spaces and arbitrary gauge functions satisfying the doubling condition.

(iv) Finally, the problem for packing measures was solved in [Joyce and Preiss (1995)] for Suslin sets A.

The proofs in (iii) and (iv) are rather long and exploit advanced techniques from functional analysis. Therefore we will omit them. For $X = \mathbb{R}^n$ the classical approach is based on *net measures* comparable with Hausdorff measures. For interested readers a version of this will be demonstrated in Theorem 2.42.

Remark 2.41. Taking into account Theorem 1.25 for the proof of Theorem 2.40 it remains to consider the case of Borel sets A with $\mu(A) = \infty$, where $\mu = \mathcal{H}^h$ (or $\mu = \mathcal{P}^h$). Here it suffices to show that there is a compact subset K of positive and finite measure by the following arguments. Assume

$$s := \sup\{\mu(K) : K \text{ compact}, K \subset A\} < \infty.$$

Then there is an increasing sequence of compact subsets K_n of A such that

$$s - \mu(K_n) < 1/n.$$

For $A' := \bigcup_{n=1}^{\infty} K_n$ we obtain $\mu(A') = s$, hence $\mu(A \setminus A') = \infty$. Choose a compact $K' \subset A \setminus A'$ with $0 < \mu(K') < \infty$. Then $K_n \cup K'$ is also a compact subset of A and we have

$$\mu(K_n \cup K') = \mu(K_n) + \mu(K')$$

which converges to $s + \mu(K')$ as $n \to \infty$. This is a contradiction to the supremum property of s.

We will prove the following special statement:

Theorem 2.42. *For any closed set $A \subset \mathbb{R}^n$ with $\mathcal{H}^h(A) > 0$ there is some compact $K \subset A$ such that $0 < \mathcal{H}^h(K) < \infty$, provided that h satisfies the doubling condition.*

Proof. We use the complete induction in n. For $n = 0$ there is nothing to prove. Suppose that the assumption holds in $\mathbb{R}^{n-1}$.

Step 1. An auxiliary family of finite net measures $(\nu_i)_{i=1}^{\infty}$ is introduced as follows: $\mathcal{C}_0 := \{C\}$, where C is an n-dimensional compact cube such that $\mathcal{H}^h(A \cap C) > 0$. For $j \geq 1$ let $\mathcal{C}_j$ be the family of closed cubes arising from $\mathcal{C}_{j-1}$ by subdividing each of the available cubes into 2^n component cubes.

Considering the cube family $\mathcal{C}_i' := \bigcup_{j=i}^{\infty} \mathcal{C}_j$ as a covering system and choosing $\tau(D) := h(|D|)$, $D \neq \emptyset$, in Carathéodory's covering construction 1.27, the resulting outer measure $\mu_{\sqrt{n}2^{-i}}^{\tau}$ will be denoted by ν_i. Then we have for any $E \subset C$,

$$\mathcal{H}_{\sqrt{n}2^{-i}}^h(E) \leq \nu_i(E) \leq \text{const } \mathcal{H}_{2^{-i}}^h(E). \tag{2.11}$$

The first inequality is obvious since the covering system for the Hausdorff measure is larger. In order to prove the second inequality consider any 2^{-i}-covering $(S_k)_{k=1}^{\infty}$ of E. If

$$2^{-(j_k+1)} \leq |S_k| < 2^{-j_k},$$

then S_k intersects at most 2^n of the cubes from $\mathcal{C}_{j_k}$ and $\sqrt{n}2^{-j_k} \leq 2\sqrt{n}|S_k|$. Hence, for the special choice of S_k such that

$$\mathcal{H}_{2^{-i}}^h(E) \geq 2^n \sum_{k=1}^{\infty} h(|S_k|) - \varepsilon$$

we obtain, in view of the doubling condition on h,

$$\mathcal{H}_{2^{-i}}^h(E) \geq \text{const} \sum_{k=1}^{\infty} h(2\sqrt{n}|S_k|) - \varepsilon$$

$$\geq \text{const } 2^n \sum_{k=1}^{\infty} h(\sqrt{n}2^{-j_k}) - \varepsilon.$$

Note that $\sqrt{n}2^{-j_k}$ is the diameter of a cube from $\mathcal{C}_{j_k}$. Therefore the last expression is not less than const $\nu_i(E) - \varepsilon$. Letting $\varepsilon \to 0$ we get the above assertion.

Step 2. If there is some hyperplane H_{n-1} such that $\mathcal{H}^h(A \cap H_{n-1}) > 0$, then we are in the situation of $\mathbb{R}^{n-1}$ and the asserted set K exists. Otherwise, $\mathcal{H}^h(A \cap H_{n-1}) = 0$ and, by (2.11),

$$\nu_i(A \cap H_{n-1}) = 0 \quad \text{for all hyperplanes } H_{n-1}. \tag{2.12}$$

Step 3. For any $E \subset A \cap C$ and $k \leq i$ we have

$$\nu_i(E) = \sum_{W \in \mathcal{C}_k} \nu_i(E \cap W). \tag{2.13}$$

To see this first note that

$$\sum_{W \in \mathcal{C}_k} \nu_i(E \cap W) \overset{(2.12)}{=} \sum_{W \in \mathcal{C}_k} \nu_i(E \cap \overset{\circ}{W})$$

$$\leq \sum_{W \in \mathcal{C}_k} \sum_{\widetilde{W} \in \widetilde{\mathcal{C}}_i, \widetilde{W} \subset W} h(|\widetilde{W}|) = \sum_{\widetilde{W} \in \widetilde{\mathcal{C}}_i} h(|\widetilde{W}|),$$

where $\overset{\circ}{W}$ is the interior of the cube W and $\widetilde{\mathcal{C}}_i$ is an arbitrary covering of E by cubes from the above system $\mathcal{C}'_i$. Thus, $\sum_{W \in \mathcal{C}_k} \nu_i(E \cap W) \leq \nu_i(E)$. The opposite inequality is obvious.

Step 4. If $E \subset A \cap C$ and $\nu_{i+1}(E \cap W) \leq h(|W|)$ for any $W \in \mathcal{C}_i$, then we get

$$\nu_{i+1}(E) = \nu_i(E). \tag{2.14}$$

In fact $\nu_{i+1}(E \cap W) = \nu_i(E \cap W)$ for any $W \in \mathcal{C}_i$, since either W provides an optimal $\mathcal{C}'_i$-covering of $E \cap W$, i.e., $h(|W|) = \nu_i(E \cap W) \leq \nu_{i+1}(E \cap W) \leq h(|W|)$, or it is sufficient to consider $\mathcal{C}'_{j+1}$-coverings of $E \cap W$. Then (2.13) implies

$$\nu_{i+1}(E) = \sum_{W \in \mathcal{C}_i} \nu_{i+1}(E \cap W) = \sum_{W \in \mathcal{C}_i} \nu_i(E \cap W) = \nu_i(E).$$

Step 5. If D is a compact subset of $A \cap C$ and $0 \leq y \leq \nu_j(D)$, then there exists a compact $E \subset D$ such that $\nu_j(E) = y$. To show this we observe that

$$f(t) := \nu_j(D \cap \{x = (x_1, \ldots, x_n) : x_1 \leq t\})$$

is a continuous function in t by the following arguments: First note that $\mathcal{H}^h(D \cap \{x : x_1 = t\}) = 0$, thanks to *Step 2*. Hence, for any $\varepsilon > 0$ there exists an open $\sqrt{n}2^{-j}$-covering $(G_k)_{k=1}^{\infty}$ of $D \cap \{x : x_1 = t\}$ such that $\sum_{k=1}^{\infty} h(|G_k|) < \varepsilon$. For sufficiently small $\delta > 0$ this is also a covering of the set $D \cap \{x : t - \delta < x_1 < t + \delta\}$ which implies

$$\mathcal{H}^h_{\sqrt{n}2^{-j}}(D \cap \{x : t - \delta < x_1 < t + \delta\}) < \varepsilon.$$

Therefore the right-hand side of (2.11) leads to the desired continuity of $f(t)$. By monotonicity f takes any value between 0 and the upper bound.

Step 6. We now will construct the desired set K. By $\mathcal{H}^h(A \cap C) > 0$ and the left-hand side of (2.11) there exists an $l \in \mathbb{N}$ with

$$\nu_l(A \cap C) > 0 \quad \text{and} \quad \nu_i(A \cap H_{n-1}) = 0, \quad i \leq l,$$

for any hyperplane H_{n-1}. Then we define inductively compact sets

$$A \cap C =: E_l \supset E_{l+1} \supset E_{l+2} \cdots,$$

such that

$$\nu_j(E_j \cap W) = \nu_{j-1}(E_{j-1} \cap W), \quad W \in \mathcal{C}_{j-1}, \quad j > l. \tag{2.15}$$

E_j arises from E_{j-1} as follows: For any $W \in \mathcal{C}_{j-1}$ choose a compact set E_W according to *Step 5* such that

$$E_W \subset E_{j-1} \cap W =: D \quad \text{and} \quad \nu_j(E_W) = \nu_{j-1}(E_{j-1} \cap W) =: y$$

(this is possible since $y = \nu_{j-1}(E_{j-1} \cap W) \leq \nu_j(E_{j-1} \cap W)$, for $W \in \mathcal{C}_{j-1}$). Put

$$E_j := \bigcup_{W \in \mathcal{C}_{j-1}} E_W$$

(then $E_j \cap W = E_W$, which yields (2.15)). From (2.13) and (2.15) we infer

$$\nu_j(E_j) = \nu_{j-1}(E_{j-1}), \quad j > l. \tag{2.16}$$

If $l \leq i < j$ we get by monotonicity of ν_{i+1} that

$$\nu_{i+1}(E_j \cap W) \leq \nu_{i+1}(E_{i+1} \cap W) \overset{(2.15)}{=} \nu_i(E_i \cap W) \leq h(|W|),$$

if $W \in \mathcal{C}_i$. Hence, (2.14) implies

$$\nu_i(E_j) = \nu_{i+1}(E_j), \quad l \leq i < j. \tag{2.17}$$

Finally (2.16) and (2.17) lead to

$$\nu_i(E_j) = \nu_l(E_l), \quad l \leq i \leq j. \tag{2.18}$$

We now set

$$K := \bigcap_{j=l}^{\infty} E_j.$$

Step 7. It remains to check the desired properties of the compact set K. By the left-hand side of (2.11) and (2.18),

$$\mathcal{H}^h(K) \leq \lim_{i \to \infty} \nu_i(K) \leq \lim_{i \to \infty} \nu_i(E_l) = \nu_l(E_l) < \infty.$$

On the other hand,

$$\mathcal{H}^h(K) \geq \mathcal{H}^h_{2^{-(l+1)}}(K) \geq \lim_{j \to \infty} \mathcal{H}^h_{2^{-l}}(E_j)$$

$$\geq \text{const} \lim_{j \to \infty} \nu_l(E_j) = \text{const} \, \nu_l(E_l) > 0,$$

since $E_l = A \cap C$. The first inequality follows from the definition of $\mathcal{H}^h$ through $\mathcal{H}^h_\delta$, the third inequality from the right-hand side of (2.11), and the equality from (2.18). It remains to prove the second inequality: We have

$$\mathcal{H}^h_{2^{-l}}(E_j) \leq \mathcal{H}^{h,\mathcal{G}}_{2^{-l}}(E_j) \leq \mathcal{H}^{h,\mathcal{G}}_{2^{-l}}(K)$$

for sufficiently large j such that any open 2^{-l}-covering of K also provides one of E_j. As in the proof of Theorem 2.2 we obtain

$$\mathcal{H}^{h,\mathcal{G}}_{2^{-l}}(K) \leq \mathcal{H}^h_{2^{-(l+1)}}(K).$$

Thus, $\lim_{j \to \infty} \mathcal{H}^h_{2^{-l}}(E_j) \leq \mathcal{H}^h_{2^{-(l+1)}}(K)$. $\qquad \square$

Chapter 3

Upper and Lower Densities of Measures and Comparison with Hausdorff and Packing Measures

3.1 Fractional densities

Extremely useful local characteristics of (outer) *Borel measures* μ in a metric space (X, d) are their upper and lower densities. Let h be a gauge function as used in the definition of Hausdorff or packing measure such that

$$\lim_{u \to 0} h(u) = 0 \,.$$

Definition 3.1.

$$\overline{\mathcal{D}}^h \mu(x) := \limsup_{r \to 0} \frac{\mu(B(x, r))}{h(2r)}$$

$$\underline{\mathcal{D}}^h \mu(x) := \liminf_{r \to 0} \frac{\mu(B(x, r))}{h(2r)}$$

are called the *upper*, resp. *lower*, *h-density* of μ at $x \in X$. For $h(u) = u^s$ we write $\overline{\mathcal{D}}^s \mu(x)$ and $\underline{\mathcal{D}}^s \mu(x)$, respectively. The *h-density* $\mathcal{D}^h \mu(x)$ exists, if $\overline{\mathcal{D}}^h \mu(x) = \underline{\mathcal{D}}^h \mu(x)$, and coincides then with this common value.

In general, we have $0 \le \underline{\mathcal{D}}^h \mu(x) \le \overline{\mathcal{D}}^h \mu(x) \le \infty$.

Example 3.2. Consider the middle–third Cantor set $C = C(1/3)$ from 2.37 and take $\mu := \mathcal{H}^D {\llcorner} C$, with $D = \log 2 / \log 3$. Our aim is to show that

$$\underline{\mathcal{D}}^D \mu(x) < \overline{\mathcal{D}}^D \mu(x), \qquad x \in C \,.$$

(In Chapter 7 it will be clear in a more general context that $\underline{\mathcal{D}}^D \mu(x) > 0$ and $\overline{\mathcal{D}}^D \mu(x) < \infty$ at any $x \in C$.)

For simplicity we will use the fact that $\mu(C) = 1$. Note that μ possesses the self-similarity property

$$\mu = \frac{1}{2}\mu \circ S_1^{-1} + \frac{1}{2}\mu \circ S_2^{-1}$$

which was already shown in 2.37. Iterating this and taking into regard that μ has no mass on the cutouts of length 3^{-n} in the n-th step of the construction of C, we obtain for any $x \in C$,

$$\mu(B(x, 3^{-n})) = 2^{-n}.$$

Therefore

$$\frac{\mu(B(x, 3^{-n}))}{(2\,3^{-n})^D} = \frac{2^{-n}}{3^{-nD}}2^{-D} = 2^{-D}$$

for any n, i.e.,

$$\underline{\mathcal{D}}^D \mu(x) \le 2^{-D}, \quad x \in C.$$

Let $C_Q \subset C$ be the (countable) set of endpoints of the intervals of C_n in the iterated construction steps. (Note that $\mu(C_Q) = 0$). The elements x of $C \setminus C_Q$ are exactly those points which lie in the images of the primary interval under infinitely many changes of the mappings S_1 and S_2. Consequently, for $x \in C \setminus C_Q$ there exists a sequence $(n_k)_{k=1}^{\infty}$ such that the maximal distance $\delta_{n_k}(x)$ from x to the endpoints of the interval I_{n_k} in the n_k-th step containing x does not exceed

$$\frac{2}{3^{n_k+1}} + \frac{1}{3^{n_k+2}} = \frac{7}{3^{n_k+2}}.$$

Then we get

$$\frac{\mu(B(x, \delta_{n_k}))}{(2\delta_{n_k})^D} = \frac{3^{-n_k D}}{2^D \delta_{n_k}^D} \ge \frac{3^{-n_k D}\,9^D}{3^{-n_k D}\,7^D}2^{-D} = \left(\frac{9}{7}\right)^D 2^{-D},$$

i.e.,

$$\overline{\mathcal{D}}^D \mu(x) \ge \left(\frac{9}{7}\right)^D 2^{-D}, \quad x \in C \setminus C_Q.$$

Exercise 3.1. Show that $\underline{\mathcal{D}}^D \mu(x) < \overline{\mathcal{D}}^D \mu(x)$, also for $x \in C_Q$.

Remark 3.3. This example is not artificial, having in mind the following deep result of [Marstrand (1964)] and [Preiss (1987)]. Let μ be a non-zero finite Borel measure in $\mathbb{R}^n$ such that $0 < \underline{\mathcal{D}}^s \mu(x)$ and $\overline{\mathcal{D}}^s \mu(x) < \infty$. Then the density $\mathcal{D}^s \mu(x)$ exists at μ-almost all x, if and only if s is an integer, μ is absolutely continuous w.r.t. $\mathcal{H}^s$ and $\mu(\mathbb{R}^n \setminus E) = 0$ for some s-rectifiable set E.

Therefore the following notion also makes sense for μ as above and $s \geq 0$.

Definition 3.4.

$$\mathcal{D}_a^s \mu(x) := \lim_{\delta \to 0} \frac{1}{|\log \delta|} \int_\delta^1 \frac{\mu(B(x,r))}{(2r)^s} \frac{dr}{r}$$

is called *average s-density* of μ at x, provided the limit exists.

In Chapter 7, Theorem 7.34, we will show the existence for certain self-similar fractal measures with dimension s. More general gauge functions h and other averaging than by means of the logarithmic measure can also be used in this definition in order to apply it to some classes of fractal measures.

For estimating fractal dimensions we will deal now with upper and lower densities. Beginning with the work of [Frostman (1935)] up to now relationships between the upper densities and Hausdorff measures have been established as a powerful tool. Analogues for lower densities and packing measures have been studied only since the 1980'th (see, e.g. [Tricot (1982)], [Taylor and Tricot (1985)], [Saint Raymond and Tricot (1988)], [Haase (1988)], [Haase (1990)], [Cutler (1995)], [Mattila and Mauldin (1997)] and [Edgar (1997)]). We will present here extensions to general situations in metric spaces.

In the sequel we will need the following *measurability property*.

Proposition 3.5. *If μ is an (outer) Borel measure in a metric space (X, d), then $\mu(B(x,r))$ as function in (x,r) is Borel measurable on the product space $X \times [0, \infty)$ equipped with the usual metric.*

Remark 3.6. In particular, under the above conditions for fixed $r > 0$, $\mu(B(x,r))$ is Borel measurable in x. Consequently, the upper density $\overline{\mathcal{D}}^h \mu(x)$ and the lower density $\underline{\mathcal{D}}^h \mu(x)$ are Borel-measurable mappings from X to $[0, \infty]$.

Proof of Proposition 3.5. It is enough to show upper semicontinuity, i.e., for any $c > 0$ the set

$$C := \{(x,r) : \mu(B(x,r)) < c\}$$

is open.

If $C \neq \emptyset$, then for all $(x,r) \in C$ there exists $\delta > 0$ such that $\mu(B(x,r + \delta)) < c$, because $B(x, r + \delta) \downarrow B(x,r)$ as $\delta \to 0$. The pairs (x', r') with

$d(x, x') < \delta/2$ and $r' < r + \delta/2$ form an open neighborhood of (x, r) and we obtain for any $y \in B(x', r')$,

$$d(x, y) \leq d(x, x') + d(x', y) < \frac{\delta}{2} + r + \frac{\delta}{2} = r + \delta.$$

Hence, $B(x', r') \subset B(x, r + \delta)$ and therefore

$$\mu(B(x', r')) < c, \text{ i.e., } (r', x') \in C.$$

$\square$

3.2 Upper densities and Hausdorff measures

Denote

$$\mathbb{H}_0 := \{h \in \mathbb{H} : h(0) = 0\}.$$

The following result is fundamental when dealing with Hausdorff dimensions.

Theorem 3.7. *Let μ be Borel measure in (X, d) and let $h \in \mathbb{H}_0$ satisfy the doubling condition $h(2u) \leq Kh(u)$, $u > 0$. Then we have the following:*

(i) $\mu(A) \leq K \sup\limits_{x \in A} \overline{\mathcal{D}}^h \mu(x) \, \mathcal{H}^h(A)$, $A \in \mathfrak{X}_\mu$, *if the supremum is finite*.

(ii) $\mu(A) \geq \inf\limits_{x \in A} \overline{\mathcal{D}}^h \mu(x) \mathcal{H}^h(A)$, $A \in \mathfrak{B}(X)$, *if the infimum is positive*.

(iii) *The inequality in* (ii) *holds for any $A \subset X$, if μ is Borel regular.*

Proof. (i) It suffices to consider the case $\overline{\mathcal{D}}^h \mu(x) < b$, $x \in A$, for some constant b. Let $(G_i)_{i=1}^\infty$ be a δ-covering of A by open sets. Denote

$$A(\delta) := \{x \in A : \mu(B(x, r)) \leq b\, h(2r), \, 0 < r < \delta\}.$$

Then $A(\delta) \uparrow A$ as $\delta \to 0$. Choose for any G_i intersecting the set $A(\delta)$ some $x_i \in G_i \cap A(\delta)$. Clearly, $G_i \subset B(x_i, |G_i|)$, which yields

$$\mu(A(\delta)) \leq \sum_{G_i \cap A(\delta) \neq \emptyset} \mu(B(x_i, |G_i|))$$

$$\leq b \sum_{G_i \cap A(\delta) \neq \emptyset} h(2|G_i|) \leq b\,K \sum_{G_i \cap A(\delta) \neq \emptyset} h(|G_i|).$$

Taking the infimum over all coverings as above we infer $\mu(A(\delta)) \leq b\,K\mathcal{H}^h(A)$. By Proposition 3.5 and monotonicity of h in r the set $A(\delta)$ is μ-measurable, so that lower continuity of μ (cf. Theorem 1.24 (ii)) implies $\lim_{\delta \to 0} \mu(A(\delta)) = \mu(A)$, i.e.,

$$\mu(A) \leq b\,K\mathcal{H}^h(A).$$

(ii) and (iii) Suppose that $\underline{\mathcal{D}}^h \mu(x) > a > 0$, $x \in A$. We will show that in this case $\mu(G) \geq a \mathcal{H}^h(A)$ for any open G containing A. Then

$$\mu(A) \geq a \mathcal{H}^h(A)$$

follows from Theorem 1.24.

We may assume that $\mu(G) < \infty$. Fix $\delta \geq 0$. By the above assumptions the family

$$\mathcal{V}(\delta) := \{V \subset G : \mu(V) > a\,h(|V|),\ |V| < \delta,\ V \text{ closed}\}$$

forms a Vitali system for A, since it contains arbitrary small closed balls centred at the points of A. From Theorem 2.14 (i) we conclude the existence of a disjoint subsequence $(V_i)_i$ such that

$$\mathcal{H}^h_\delta\Big(A \setminus \bigcup_i V_i\Big) = 0,$$

taking into regard that

$$a \sum_i h(|V_i|) \leq \sum_i \mu(V_i) = \mu\Big(\bigcup_i V_i\Big) \leq \mu(G) < \infty$$

by definition of the V_i. Hence,

$$\mathcal{H}^h_\delta(A) \leq \mathcal{H}^h_\delta\Big(A \cap \bigcup_i V_i\Big) + 0 \leq \sum_i \mathcal{H}^h_\delta(V_i) \leq \sum_i h(|V_i|) \leq a^{-1}\mu(G).$$

Letting $\delta \to 0$ we infer the desired inequality $a\mathcal{H}^h(A) \leq \mu(G)$. $\qquad\square$

Remark 3.8. As a corollary of Theorem 3.7 one obtains the so-called *mass distribution principle* for estimating the Hausdorff dimension:

If the above conditions are fulfilled with upper s-densities $\overline{\mathcal{D}}^s \mu(x)$ bounded from above or from below by a positive constant and $0 < \mu(A) < \infty$, then we obtain in (i) that $\mathcal{H}^s(A) > 0$, i.e., $\dim_H A \geq s$, and in (ii) and (iii) that $\mathcal{H}^s(A) < \infty$, i.e., $\dim_H A \leq s$, respectively. (An example will be treated in 5.20, see also Chapter 7.)

In particular, if μ is a Borel measure, A is a Borel set with $0 < \mu(A) < \infty$ and

$$a \leq \overline{\mathcal{D}}^D \mu(x) \leq b\,, \quad x \in A\,,$$

for some constants $0 < a \leq b$ and $D \geq 0$, then we get

$$\dim_H A = D \text{ and } \mu \llcorner A \sim \mathcal{H}^D \llcorner A \text{ as measures on } \mathfrak{B}(X)\,.$$

(The latter follows from $\mu(B) \leq 2^D b\, \mathcal{H}^D(B) \leq 2^D ba^{-1}\mu(B)$ for any Borel set $B \subset A$.)

These density estimates are fulfilled if μ is a *D-measure*, i.e., if

$$a \leq \frac{\mu(B(x,r))}{(2r)^D} \leq b\,, x \in A, r \leq r_0\,,$$

for some constant $r_0 > 0$. A set $A \subset X$ admitting such a D-measure μ is called a *D-set*.

Recall that in Example 3.2 we have got the estimation

$$\overline{\mathcal{D}}^D \mu(x) \leq 2^{-D},$$

where μ is the D-dimensional Hausdorff measure restricted to the middle-third Cantor set, with Hausdorff dimension $D = \log 2 / \log 3$. 2^{-D} is at the same time a (sharp) lower bound for the upper density. More generally, the following *Density theorem for Hausdorff measures* is a consequence of Theorem 3.7.

Theorem 3.9. *Let $h \in \mathbb{H}_0$ satisfy the doubling condition with constant K and B be a Borel subset of the metric space X with $0 < \mathcal{H}^h(B) < \infty$. Then we have*

(i) $K^{-1} \leq \overline{\mathcal{D}}^h(\mathcal{H}^h \llcorner B)(x) \leq 1$ *at $\mathcal{H}^h - a.a.\ x \in B$,*

(ii) $\overline{\mathcal{D}}^h(\mathcal{H}^h \llcorner B)(x) = 0$ *at $\mathcal{H}^h - a.a.\ x \in B^c$.*

Proof. Note that $\mu := \mathcal{H}^h \llcorner B$ is a finite Borel measure. First choose $b < K^{-1}$ and set

$$A := \{x \in B : \overline{\mathcal{D}}^h \mu(x) \leq b\}.$$

From Theorem 3.7 (i) we get

$$\mathcal{H}^h(A) = \mu(A) \leq b K \mathcal{H}^h(A).$$

Since $bK < 1$ this yields $\mathcal{H}^h(A) = 0$, i.e., the left-hand side of assertion (i). Similarly, for $a > 1$ and

$$A := \{x \in B : \overline{\mathcal{D}}^h \mu(x) \geq a\}$$

we obtain by the measurability of A (Proposition 3.5) from Theorem 3.7 (ii)),

$$\mathcal{H}^h(A) = \mu(A) \geq a \mathcal{H}^h(A),$$

i.e., $\mathcal{H}^h(A) = 0$, which yields the right-hand side of (i).
Finally, we choose for $a > 0$,

$$A := \{x \in B^c : \overline{\mathcal{D}}^h \mu(x) \geq a\}$$

and infer from Theorem 3.7 (ii),

$$0 = \mu(A) \geq a \mathcal{H}^h(A),$$

i.e., $\mathcal{H}^h(A) = 0$ which implies assertion (ii). (In all cases we have used that A is a Borel set, which follows from Remark 3.6.) $\square$

3.3   Lower densities and packing measures

For packing measures the analogue to the comparison theorem 3.7 reads as follows. Recall that $\mathcal{P}^h \leq \widetilde{\mathcal{P}}^h$, where $\mathcal{P}^h$ was constructed by metric ball packings and $\widetilde{\mathcal{P}}^h$ by ordinary ball packings. Denote

$$\mathbb{H}^0_- := \{h \in \mathbb{H}_- : \lim_{u \to 0} h(u) = 0\}.$$

Theorem 3.10. *Let μ be a Borel measure on (X, d) and $h \in \mathbb{H}^0_-$.*

(i) *If $\dfrac{h(u)}{h(\varrho u)} \leq K + \alpha(\varrho)$, where $\lim\limits_{\varrho \uparrow 1/2} \alpha(\varrho) = 0$, then*

$$\mu(A) \leq K \sup_{x \in A} \underline{\mathcal{D}}^h \mu(x) \, \mathcal{P}^h(A), \quad A \subset X \text{ if the supremum is finite.}$$

(ii) *$\mu(A) \geq \inf\limits_{x \in A} \underline{\mathcal{D}}^h \mu(x) \, \widetilde{\mathcal{P}}^h(A), \quad A \in \mathfrak{B}(X)$, if the infimum is positive.*

Proof. (i) Suppose $\underline{\mathcal{D}}^h \mu(x) < b$, $x \in A$. It is enough to show that in this case

$$\mu(E) \leq b \, K \mathcal{P}^h_0(E), \quad E \subset A,$$

since this implies

$$\mu(A) \leq \inf \left\{ \sum_{i=1}^{\infty} \mu(A_i) : A = \bigcup_{i=1}^{\infty} A_i \right\}$$

$$\leq b \, K \inf \left\{ \sum_{i=1}^{\infty} \mathcal{P}^h_0(A_i) : A = \bigcup_{i=1}^{\infty} A_i \right\} = b \, K \mathcal{P}^h(A).$$

We may assume $\mathcal{P}^h_0(E) < \infty$, i.e., $\mathcal{P}^h_\delta(E) < \infty$ for all sufficiently small $\delta > 0$. For such δ the family

$$\mathcal{B}(\delta) := \{B(x, r) : x \in E, \ \mu(B(x, r)) \leq b \, h(2r), \ 2r < \delta\}$$

forms a centered Vitali-system of closed balls for E. By Theorem 2.30, for any $\varrho < 1/2$ there exists a δ-ball packing $(B(x_i, \varrho r_i))_i$ for E such that $B(x_i, r_i) \in \mathcal{B}(\delta)$ for all i and

$$E \subset \bigcup_i B(x_i, r_i).$$

(If infinitely many r_i are greater than some positive constant, then $\sum_{i=1}^{\infty} h(2r_i) = \infty$, which contradicts $\mathcal{P}^h_\delta(E) < \infty$.) Hence,

$$\mu(E) \leq \sum_i \mu(B(x_i, r_i)) \leq b \sum_i h(2r_i) \leq b \sum_i h(2\varrho r_i)(K + \alpha(\varrho))$$

$$\leq b(K + \alpha(\varrho))\mathcal{P}^h_\delta(E).$$

Letting $\varrho \uparrow 1/2$ and $\delta \to 0$ we obtain

$$\mu(E) \leq b\, K\mathcal{P}_0^h(E)\,,$$

which completes the proof of (i).

In order to show (ii) assume $\underline{\mathcal{D}}^h \mu(x) > a$, $x \in A$. We have to show that

$$a\widetilde{\mathcal{P}}^h(A) \leq \mu(A)\,.$$

In view of Theorem 1.24 (i) it suffices to prove

$$a\widetilde{\mathcal{P}}^h(A) \leq \mu(G)$$

for any open set G containing A. For $\delta > 0$ consider the approximating $\widetilde{\mathcal{P}}^h$-measurable sets

$$A(\delta) := \{x \in A : \mu(B(x,r)) \geq ah(2r)\,,\ B(x,r) \subset G\,,\ 2r < \delta\}\,.$$

By the assumption on the lower densities, $A(\delta) \uparrow A$ as $\delta \to 0$.

Let $(B(x_i, r_i))_i$ be an ordinary δ-ball packing of $A(\delta)$. Then we get

$$a\sum_h (2r_i) \leq \sum_i \mu(B(x_i, r_i)) = \mu\Big(\bigcup_i B(x_i, r_i)\Big) \leq \mu(G)\,,$$

since $B(x_i, r_i) \subset G$ for any i. Therefore

$$a\widetilde{\mathcal{P}}_\delta^h(A(\delta)) \leq \mu(G)\,,$$

i.e.,

$$a\widetilde{\mathcal{P}}^h(A(\delta)) \leq a\widetilde{\mathcal{P}}_0^h(A(\delta)) \leq a\widetilde{\mathcal{P}}_\delta^h(A(\delta)) \leq \mu(G)\,.$$

Finally we use that $\lim_{\delta \to 0} \widetilde{\mathcal{P}}^h(A(\delta)) = \widetilde{\mathcal{P}}^h(A)$ by monotonicity. $\square$

Remark 3.11.

(i) An analysis of the proof shows that the estimation in 3.10 (i) remains valid for any h satisfying the usual doubling condition when there K is replaced by the squared constant from the doubling condition.

Moreover, the doubling condition is not needed if μ has the so-called *strong Vitali property*: For any Borel set A and any centered Vitali system $\mathcal{B}$ of closed balls for A there exists an (ordinary) ball packing $(B(x_i, r_i))_i$ of A such that $\mu\big(A \setminus \bigcup_i B(x_i, r_i)\big) = 0$. Then the inequality in (i) may be replaced by

$$\mu(A) \leq \sup_{x \in A} \mathcal{D}^h \mu(x)\, \mathcal{P}^h(A)\,,\quad A \in \mathfrak{B}(X)$$

$$\left(\mu(A) \leq \sup_{x \in A} \mathcal{D}^h \mu(x)\, \widetilde{\mathcal{P}}^h(A)\,,\quad A \in \mathfrak{B}(X)\,,\ \text{resp.}\right).$$

It is well-known that in $\mathbb{R}^n$ every Borel measure μ has the strong Vitali property (cf. [Mattila (1995)], [Edgar (1997)]).

(ii) Packing dimension estimations follow similarly as for Hausdorff measures (cf. Remark 3.8).

The *Density theorem for packing measures* is a straightforward consequence:

Theorem 3.12. *Let B be a Borel set in (X, d) with $0 < \mathcal{P}^h(B) < \infty$ and let $h \in \mathbb{H}^0_-$. Then we have the following:*

(i) $\underline{\mathcal{D}}^h(\widetilde{\mathcal{P}}^h \llcorner B)(x) \leq 1$ *at* $\widetilde{\mathcal{P}}^h - a.a.\ x \in B$.

(ii) *If* $h(u)/h(\varrho u) \leq K + \alpha(\varrho)$ *with* $\lim_{\varrho \uparrow 1/2} \alpha(\varrho) = 0$, *then*

$$\underline{\mathcal{D}}^h(\mathcal{P}^h \llcorner B)(x) \geq K^{-1} \quad at\ \mathcal{P}^h - a.a.\ x \in B.$$

(iii) $\underline{\mathcal{D}}^h(\widetilde{\mathcal{P}}^h \llcorner B)(x) = 0$ *at* $\widetilde{\mathcal{P}}^h - a.a.\ x \in B^c$.

(iv) *If* $X = \mathbb{R}^n$ *with the usual metric then*

$$\underline{\mathcal{D}}^h(\mathcal{P}^h \llcorner B)(x) = 1 \quad at\ \mathcal{P}^h - a.a.\ x \in B.$$

Remark 3.13. As a corollary of the above Density theorems for Hausdorff and packing measures in $\mathbb{R}^n$ for $0 < \mathcal{P}^s(A) < \infty$ one obtains the equality

$$\mathcal{P}^s(A) = \mathcal{H}^s(A),$$

if and only if the density $D^s(\mathcal{H}^s \llcorner A)(x)$ exists and equals 1 for $\mathcal{H}^s$-almost all $x \in A$. According to Remark 3.3 in this case s must be an integer and A an s-rectifiable set. (Cf. [Saint Raymond and Tricot (1988)], [Mattila (1995)].)

3.4 Frostman lemma for Hausdorff measures

In the Comparison Theorem 3.7 we have seen that upper densities of Borel measures are closely related to Hausdorff measures with the same gauge function. The following statement is a certain counterpart to the first part of this theorem and the resulting mass distribution principle. From the literature it is known as Frostman lemma for the special cases of $\mathbb{R}^n$ and of compact metric spaces (cf. [Mattila (1995)], where methods of functional analysis are used, and the references therein).

Recall that the *support* of a Borel measure μ on (X, d) is defined as

$$\operatorname{supp} \mu := X \setminus \bigcup \{G : G \text{ open}, \ \mu(G) = 0\}$$

or equivalently, as the set of points for which every open neighborhood has positive measure. Furthermore, a complete separable metric space is also called *Polish space*.

Theorem 3.14. *Let (X, d) be a Polish space, $h \in \mathbb{H}_0$ satisfy the doubling condition, and A be a Borel set with $\mathcal{H}^h(A) > 0$. Then there exists a Borel regular measure μ with compact support in A such that $0 < \mu(A) < \infty$ and*

$$\frac{\mu(B(x, r))}{h(2r)} \leq c, \quad x \in A, \ r > 0, \ \text{for some constant } c > 0.$$

Proof. By Theorem 2.40 there exists a compact set $K \subset A$ such that $0 < \mathcal{H}^h(K) < \infty$. The Density theorem for Hausdorff measures (Corollary 3.9) implies $\overline{\mathcal{D}}^h(\mathcal{H}^h \llcorner K)(x) \leq 1$ at $\mathcal{H}^h$-a.a. $x \in K$. Then there must be some $c_0 > 1$ and $r_0 > 0$ such that for the set

$$K_0 := \{x \in K : \mathcal{H}^h(K \cap B(x, r)) \leq c_0 h(2r), \ 0 < r < r_0\}$$

we have

$$0 < \mathcal{H}^h(K_0) < \infty.$$

K_0 is a closed subset of the compact set K, and therefore it is compact, too. (Closeness follows from $B(x, r) \subset B(y, r + \varepsilon)$, if $d(x, y) < \varepsilon$, together with right continuity of h.) For $r \geq r_0$ we can estimate

$$\frac{\mathcal{H}^h(K \cap B(x, r))}{h(2r)} \leq \frac{\mathcal{H}^h(K)}{h(2r_0)}.$$

Setting $c_1 := \max\{c_0, \mathcal{H}^h(K)/h(2r_0)\}$ and

$$\mu := \mathcal{H}^h \llcorner K_0$$

we obtain

$$\frac{\mu(B(x, r))}{h(2r)} = \frac{\mathcal{H}^h(K_0 \cap B(x, r))}{h(2r)} \leq \frac{\mathcal{H}^h(K \cap B(x, r))}{h(2r)} \leq c_1$$

for any $x \in K_0$, i.e., for μ-a.a. $x \in A$. If $x \in A$ is arbitrary and $\mu(B(x, r)) > 0$, then there exists some $y \in A \cap B(x, r)$ such that

$$\frac{\mu(B(y, r))}{h(2r)} \leq c_1 \ \text{for all } r.$$

Hence,

$$\frac{\mu(B(x, r))}{h(2r)} \leq \frac{\mu(B(y, 2r))}{h(2r)} \leq \frac{c_1 h(4r)}{h(2r)} \leq \text{const}$$

by the doubling condition on h. $\qquad\square$

Chapter 4

Hausdorff Dimension and Potential Theory

The mass distribution principle 3.8 for estimating the Hausdorff dimension and the Frostman lemma 3.14 can also be used in a potential theoretic context. In this chapter we will show that in Polish spaces (X, d) the Hausdorff dimension of Borel sets can be interpreted as capacity dimension. Moreover, we present the corresponding *potential theoretic method for estimating the Hausdorff dimension*. (For the Euclidean case, which goes back to [Frostman (1935)], cf. [Falconer (1990)] with the examples and references therein.)

In analogy to physical concepts the following notions for a *mass distribution*, i.e., a non-zero finite Borel measure μ in a metric space (X, d), are introduced:

Definition 4.1. For a Borel measure μ on (X, d) with $0 < \mu(X) < \infty$ and $s \geq 0$,

$$U_\mu^s(x) := \int d(x, y)^{-s} \, \mu(dy)$$

is called *s-potential* of μ at point $x \in X$, and

$$E^s(\mu) := \int \int d(x, y)^{-s} \, \mu(dy)\mu(dx)$$

$$\left(= \int U_\mu^s(x) \, \mu(dx) \right)$$

is the *s-energy* of μ.

For $A \subset X$ denote

$$\mathcal{M}(A) := \{\mu : \mu \text{ Borel measure on } X, \ \operatorname{supp} \mu \subset A, \ \mu(X) = 1\}. \quad (4.1)$$

Definition 4.2.

$$C^s(A) := \sup\left\{(E_\mu^s)^{-1} : \mu \in \mathcal{M}(A)\right\}$$

is called *s-capacity* of the set $A \subset X$.

Exercise 4.1. C^s is an outer measures on X.

In general, C^s is no Borel measure. For $X = \mathbb{R}^n$ an example with references to the literature can be found in [Mattila (1995)] [Remarks 8.6].

Note that $C^s(A) > 0$ if and only if there is some $\mu \in \mathcal{M}(A)$ with $E_\mu^s < \infty$.

Below we will prove the following basic relationship between s-capacities and Hausdorff measures.

Theorem 4.3.

   (i) $\mathcal{H}^s(A) < \infty$ *implies* $C^s(A) = 0$.
   (ii) $C^s(A) = 0$ *implies* $\mathcal{H}^t(A) = 0$, $t > s$,
        provided (X, d) *is a Polish space and* $A \in \mathfrak{B}(X)$.

Thus, the corresponding dimension notion makes sense:

Definition 4.4.

$$\dim_C A := \sup\{s : C^s(A) > 0\} = \inf\{s : C^s(A) = 0\}$$

is called *capacity dimension* of the Borel set A in the Polish space (X, d).

From Theorem 4.3 we immediately infer that *the Hausdorff dimension is equal to the capacity dimension:*

Corollary 4.5. *If* (X, d) *is a Polish space and* $A \in \mathfrak{B}(X)$, *then*

$$\dim_H A = \dim_C A.$$

Remark 4.6.

   (i) The outer measure C^s has more zero-sets than $\mathcal{H}^s$. (For example, for a rectifiable curve γ with $0 < \mathcal{H}^1(\gamma) < \infty$ it can be shown that $C^1(\gamma) = 0$.)
   (ii) The finiteness of C^s provides less information than that of $\mathcal{H}^s$. ($C^s(B) < \infty$ for any bounded Borel set B by a simple estimate of the s-energy of $\mu \in \mathcal{M}(B)$:
   $E_\mu^s = \int_B \int_B d(x, y)^{-s}\, \mu(dy)\mu(dx) \geq |B|^{-s}\mu(B)^2 = |B|^{-s}$.)

Nevertheless, C^s is rather useful for *lower dimension estimates* (cf. 3.8). An immediate consequence of Theorem 4.3 is the following. (Applications will be demonstrated in the next chapters.)

Corollary 4.7. *Let A be a Borel set in* (X, d).

(i) *If there is some* $\mu \in \mathcal{M}(A)$ *with*

$$\int \int d(x, y)^{-s} \mu(dy)\mu(dx) < \infty,$$

then $\dim_H A \geq s$.

(ii) *If* (X, d) *is a Polish space and* $\dim_H A > s$, *then there exists a measure* μ *as in* (ii).

The first part of 4.7 can be improved by requiring only the finiteness of the s-potentials at almost all x.

Theorem 4.8. *If* $A \in \mathcal{B}(X)$ *and there is some* $\mu \in \mathcal{M}(A)$ *such that*

$$\int d(x, y)^{-s} \mu(dy) < \infty \quad at \ \mu - a.a. \ x \in A,$$

then $\dim_H A \geq s$.

This version is extremely useful in probabilistic applications.

Proof of Theorems 4.3 and Theorem 4.8. We first show Theorem 4.3 (i) and Theorem 4.8:

Assume $C^s(A) > 0$, i.e., the contrary of assertion (i). Then there is some $\mu \in \mathcal{M}(A)$ with

$$\int \int d(x, y)^{-s} \mu(dy)\mu(dx) < \infty.$$

Hence,

$$\int d(x, y)^{-s} \mu(dy) < \infty \quad at \ \mu - a.a. \ x \in A,$$

and therefore the condition of Theorem 4.8 is fulfilled. For this case there exists a Borel set $A_0 \subset A$ such that $\mu(A_0) > 0$ and the last inequality holds for all $x \in A_0$. At these x we get

$$\lim_{r \to 0} \int_{B(x,r)} d(x, y)^{-s} \mu(dy) = 0$$

and consequently,

$$\limsup_{r \to 0} \frac{\mu(B(x,r))}{r^s} \leq \limsup_{r \to 0} \int_{B(x,r)} d(x, y)^{-s}\mu(dy) = 0,$$

and thus, $\overline{\mathcal{D}}\mu(x) = 0$, $x \in A_0$. By the Density comparison theorem for Hausdorff measures 3.7 (i), i.e., by

$$\mu(A_0) \leq \text{const} \sup_{x \in A_0} \overline{\mathcal{D}}^s \mu(x) \, \mathcal{H}^s(A_0) ,$$

we conclude that $\infty = \mathcal{H}^s(A_0) \leq \mathcal{H}^s(A)$. This is a contradiction to $\mathcal{H}^s(A) < \infty$, and we infer

$$C^s(A) = 0 ,$$

which completes the proof of 4.3 (i). Moreover, the above arguments have shown that under the conditions of Theorem 4.8 we get $\mathcal{H}^s(A) = \infty$, i.e., $\dim_H(A) \geq s$, which proves the assertion of this theorem.

It remains to show Theorem 4.3 (ii): Assume the contrary, i.e., $\mathcal{H}^t(A) > 0$ for some $t > s$. Then we will show that there is a measure $\mu \in \mathcal{M}(A)$ with $E_\mu^s < \infty$, i.e., $C^s(A) > 0$, which yields a contradiction.

By the Frostman Lemma 3.14 there exists a $\mu \in \mathcal{M}(A)$ satisfying $\mu(B(x,r)) \leq cr^t$, $r > 0$, for $x \in A$ and some constant $c > 0$. We now consider $m_x(r) := \mu(B(x,r))$ as distribution function of a probability measure on $(0,+\infty)$ and decompose

$$U_\mu^s(x) = \int d(x,y)^{-s} \, \mu(dy) = \int_0^\infty r^{-s} \, m_x(dr)$$

$$= \int_0^1 r^{-s} \, m_x(dr) + \int_1^\infty r^{-s} \, m_x(dr)$$

$$\leq \int_0^1 r^{-s} \, m_x(dr) + 1$$

$$= r^{-s} m_x(r) \big|_0^1 + s \int_0^1 r^{-s-1} \, m_x(r) \, dr + 1$$

(the partial integration follows from Fubini). By the above estimate of $m_x(r) = \mu(B(x,r))$ the last expression does nor exceed

$$1 - \lim_{r \to 0} r^{-s} m_x(r) + s \int_0^1 r^{-s-1} cr^t \, dr + 1$$

$$= 1 - 0 + \frac{cs}{t-s} r^{t-s} \big|_0^1 + 1 = 2 + \frac{cs}{t-s} .$$

Therefore

$$U_\mu^s(x) \leq 2 + \frac{cs}{t-s} , \quad x \in A , \quad \text{and thus,} \quad E_\mu^s \leq 2 + \frac{cs}{t-s} < \infty ,$$

which completes the proof. $\square$

Chapter 5

Other Fractal Dimensions

For practical purposes Hausdorff and packing dimensions are sometimes difficult to handle, because of their fine constructions. Often it is more convenient to work with coverings or packings of equal size. This approach will be made precise in this section in form of entropy, metric and box dimensions. Closely related are dimension notions in $\mathbb{R}^n$ basing on volumes of parallel sets which we start with.

5.1 Minkowski dimension in Euclidean spaces

Recall that $d(x, A) := \inf\{d(x, y) : y \in A\}$ denotes the distance of a point x to the set A in the metric space (X, d), and for $\delta \geq 0$,

$$A_\delta := \{x \in X : d(x, A) \leq \delta\}$$

is the δ-parallel set of A.

Definition 5.1. For $s \geq 0$ the *upper* and *lower s-dimensional Minkowski content* of $A \subset \mathbb{R}^n$ are defined by

$$\overline{\mathcal{M}}^s(A) := \limsup_{\delta \to 0} \frac{\mathcal{L}^n(A_\delta)}{\delta^{n-s}},$$

$$\underline{\mathcal{M}}^s(A) := \liminf_{\delta \to 0} \frac{\mathcal{L}^n(A_\delta)}{\delta^{n-s}},$$

respectively. If the limits agree, then the common value $\mathcal{M}^s(A)$ is called *s-dimensional Minkowski content*.

Remark 5.2. It can be shown that the Minkowski contents are not σ-subadditive. For m-dimensional C^1-submanifolds M_m of $\mathbb{R}^n$ it is known that

$$\mathcal{M}^m(M_m) = \mathcal{H}^m(M_m).$$

$\mathcal{M}^s$ and the related dimension notion play an important role in the theory of partial differential equations on domains of $\mathbb{R}^n$ with fractal boundary. The corresponding dimensions will be introduced similarly as Hausdorff and packing dimensions.

In the sequel for brevity we will use the notation $\overline{(\cdot)}$ for either $\overline{(\cdot)}$, i.e., upper, or $\underline{(\cdot)}$, i.e., lower notions under consideration.

Proposition 5.3. $\underline{\mathcal{M}}^s(A) < \infty$ *implies* $\underline{\mathcal{M}}^t(A) = 0$ *for all* $t > s$.

Proof. If $\underline{\mathcal{M}}^s(A) < \infty$, then there exists a sequence $\delta_k \to 0$ as $k \to \infty$, such that

$$\frac{\mathcal{L}^n(A_{\delta_k})}{\delta_k^{n-s}} < \text{const}.$$

For $t > s$ one infers

$$\frac{\mathcal{L}^n(A_{\delta_k})}{\delta_k^{n-t}} = \frac{\mathcal{L}^n(A_{\delta_k})}{\delta_k^{n-s}} \, \delta_k^{t-s},$$

which tends to 0 as $k \to \infty$, i.e.,

$$\liminf_{\delta \to 0} \frac{\mathcal{L}^n(A_\delta)}{\delta^{n-t}} = 0.$$

The proof for $\overline{\mathcal{M}}^s$ is similar. □

Thus, we may introduce the following.

Definition 5.4.

$$\overline{\dim}_M A := \sup\{s \geq 0 : \overline{\mathcal{M}}^s(A) = \infty\} = \inf\{s \geq 0 : \overline{\mathcal{M}}^s(A) = 0\},$$

$$\underline{\dim}_M A := \sup\{s \geq 0 : \underline{\mathcal{M}}^s(A) = \infty\} = \inf\{s \geq 0 : \underline{\mathcal{M}}^s(A) = 0\}$$

are called *upper* and *lower Minkowski dimensions* of $A \subset \mathbb{R}^n$, respectively. If $\overline{\dim}_M A = \underline{\dim}_M A$, then the common value $\dim_M A$ is the *Minkowski dimension* of the set A.

Remark 5.5. $\underline{\dim}_M A \leq \overline{\dim}_M A$.

There is a more explicit method for determining these dimensions.

Proposition 5.6.

$$\overline{\dim}_M A = n - \liminf_{\delta \to 0} \frac{\log \mathcal{L}^n(A_\delta)}{\log \delta},$$

$$\underline{\dim}_M A = n - \limsup_{\delta \to 0} \frac{\log \mathcal{L}^n(A_\delta)}{\log \delta}.$$

for any bounded set A.

(Because of the ratios instead of ln one can use there the logarithm with respect to an arbitrary basis, which will be denoted by log in the sequel.)

Proof.

$$\limsup_{\delta\to 0}\frac{\mathcal{L}^n(A_\delta)}{\delta^{n-s}} = \limsup_{\delta\to 0}\exp\left\{(\log\mathcal{L}^n(A_\delta) - (n-s)\log\delta\right\}$$

$$= \exp\left\{\limsup_{\delta\to 0}\left(n - \frac{\log\mathcal{L}^n(A_\delta)}{\log\delta} - s\right)(-\log\delta)\right\}$$

$$= \begin{cases} \infty, & \text{if } s < n - \liminf_{\delta\to 0}\frac{\log\mathcal{L}^n(A_\delta)}{\log\delta}, \\ 0, & \text{if } s > n - \liminf_{\delta\to 0}\frac{\log\mathcal{L}^n(A_\delta)}{\log\delta}. \end{cases}$$

The arguments for $\liminf$ are similar. This leads to the desired equalities. $\qquad\square$

5.2 Entropy dimension and metric dimension

Let now (X, ρ) be a metric space, $A \subset X$, and $\delta > 0$.

By $n_\delta(A)$ we denote the *minimal number of closed balls of radius δ covering* A. A is said to be *totally bounded* if $n_\delta(A) < \infty$ for any $\delta > 0$. Throughout this section we will consider such A.

The following notions go back to [Pontryagin and Shnirelman (1932)].

Definition 5.7.

$$\overline{\dim}_E A := \limsup_{\delta\to 0}\frac{\log n_\delta(A)}{|\log\delta|},$$

$$\underline{\dim}_E A := \liminf_{\delta\to 0}\frac{\log n_\delta(A)}{|\log\delta|}$$

are called *upper* and *lower entropy dimension* of A, respectively. If the limits agree then the common value $\dim_E A$ is the *entropy dimension* of A.

Exercise 5.1.

(i) $\overline{\dim}_E A$ may also be determined like the Hausdorff dimension, if instead of general δ-coverings in the construction of $\mathcal{H}_\delta^s$ only balls of the same radius δ are admitted and then the upper, resp. lower limit as $\delta \to 0$ is considered.

(ii) $\dim_H A \leq \underline{\dim}_E A \leq \overline{\dim}_E A$.

The packing counterparts of these notions are introduced as follows. Let $N_\delta(A) \le \infty$ be the *maximal number of elements of a packing for A with balls of the same radius δ.*

Definition 5.8.

$$\overline{\dim}_\varrho A := \limsup_{\delta \to 0} \frac{\log N_\delta(A)}{|\log \delta|},$$

$$\underline{\dim}_\varrho A := \liminf_{\delta \to 0} \frac{\log N_\delta(A)}{|\log \delta|}$$

are called *upper* and *lower metric dimension* of A, respectively. If the limits agree then the common value $\dim_\varrho A$ is the *metric dimension* of A.

Theorem 5.9. *For any totally bounded set A in a metric space (X, ϱ) we have*

$$\overline{\dim}_E A = \overline{\dim}_\varrho A,$$

$$\underline{\dim}_E A = \underline{\dim}_\varrho A.$$

Proof. Suppose that $B_1, \dots, B_{N_\delta(A)}$ are the elements of an optimal ordinary ball packing for A with radii δ and $N_\delta(A) < \infty$. Any $x \in A$ has a distance not greater than 2δ to some of the centers of these balls. (Otherwise $B(x, \delta)$ may be added to this packing, which contradicts the maximality of $N_\delta(A)$.) Therefore A may be covered by $N_\delta(A)$ balls of radii 2δ, i.e.,

$$n_{2\delta}(A) \le N_\delta(A)$$

which holds also, if $N_\delta(A) = \infty$. This yields

$$\frac{\log n_{2\delta}(A)}{|\log 2\delta|} = \frac{\log n_{2\delta}(A)}{|\log \delta + \log 2|} \le \frac{\log n_{2\delta}(A)}{|\log \delta|} \left|1 + \frac{\log 2}{\log \delta}\right|^{-1},$$

hence,

$$\overline{\dim}_E A = \overline{\dim}_{\delta \to 0} \frac{\log n_{2\delta}(A)}{|\log 2\delta|} \le \overline{\lim}_{\delta \to 0} \frac{\log N_\delta(A)}{|\log \delta|} = \overline{\dim}_\varrho A$$

writing $\overline{\lim}$ for the upper, resp. lower, limit.

Conversely, let $\widetilde{B}_1, \dots, \widetilde{B}_{n_{\delta/3}(A)}$ be an optimal covering of A with balls of radii $\delta/3$, $n_{\delta/3}(A) < \infty$ and B_i be from a packing for A of balls with radius δ. Then the center of B_i is contained in some $\widetilde{B}_j$. Because of the choice of the radii we get $\widetilde{B}_j \subset B_i$. Since the balls B_i are disjoint, this implies

$$N_\delta(A) \le n_{\delta/3}(A).$$

Similarly as above we now obtain

$$\overline{\dim}_\varrho A \le \overline{\dim}_E A.$$

$\square$

The following statement is in the line of the mass distribution principles for Hausdorff and packing measures.

Corollary 5.10. *If for $A \subset X$ there exist a mass distribution μ on X and constants $a, r_0 > 0$ such that*

$$\mu(B(x,r)) \geq a\, r^s, \quad x \in A, \quad r < r_0,$$

then A is totally bounded and

$$\overline{\dim}_E A = \overline{\dim}_\varrho A \leq s.$$

Proof. Let $(B(x_i, \delta))_{i=1}^k$ be a ball packing for A. Then we get for any $\delta < r_0$,

$$k\, a\delta^s \leq \sum_{i=1}^k \mu(B(x_i, \delta)) = \mu\Big(\bigcup_{i=1}^k B(x_i, \delta)\Big) \leq \mu(X) < \infty.$$

Hence,

$$N_\delta(A) \leq \text{const}\, \delta^{-s}.$$

In the proof of the last theorem we have seen that $n_{2\delta}(A) \leq N_\delta(A)$. Therefore A is totally bounded and

$$\frac{\log n_{2\delta}(A)}{|\log 2\delta|} \leq \frac{\text{const}}{|\log \delta|} + s$$

for any $\delta < r_0$. Thus,

$$\overline{\dim}_E A = \overline{\dim}_\varrho A \leq s.$$

$\square$

We now consider the special case $X = \mathbb{R}^n$ and the relationships to the corresponding Minkowski dimensions.

Theorem 5.11. *For any bounded set $A \subset \mathbb{R}^n$ we have*

$$\overline{\dim}_M A = \overline{\dim}_E A = \overline{\dim}_\varrho A,$$
$$\underline{\dim}_M A = \underline{\dim}_E A = \underline{\dim}_\varrho A.$$

Proof. The right side equalities are special cases of Theorem 5.9. Next we will estimate Minkowski dimensions by entropy dimensions. Since A has a covering of $n_\delta(A)$ balls of radius δ, the parallel sets A_δ may be covered by the same number of balls of radius 2δ. Volume estimation yields

$$\mathcal{L}^n(A_\delta) \leq n_\delta(A)\, c(n)(2\delta)^n,$$

where $c(n)$ denotes the volume of the unit ball. Thus,

$$n - \limsup_{\delta \to 0} \frac{\log \mathcal{L}^n(A_\delta)}{\log \delta} = n + \liminf_{\delta \to 0} \frac{\log \mathcal{L}^n(A_\delta)}{|\log \delta|}$$

$$\leq n + \liminf_{\delta \to 0} \frac{\log n_\delta(A) + n \log \delta + \text{const}}{|\log \delta|} = n + \underline{\dim}_E A - n.$$

Proposition 5.6 implies

$$\underline{\dim}_M A \leq \underline{\dim}_E A.$$

Analogously,

$$\overline{\dim}_M A \leq \overline{\dim}_E A.$$

Conversely, since $N_\delta(A)$ disjoint balls of radius δ have their centers in A, we obtain

$$N_\delta(A)\, c(n) \delta^n \leq \mathcal{L}^n(A_\delta)$$

and from this we get similarly as above,

$$\underline{\dim}_\varrho A \leq \overline{\dim}_M A.$$

$\square$

5.3 Box dimension in Euclidean spaces

The most practical version of the above dimensions in $\mathbb{R}^n$ arises when the balls of size δ are replaced by the boxes of an n-dimensional lattice containing the coordinate axes with lattice constant δ .

Let $N_\delta^B(A)$ be the *number of boxes which intersect the bounded set* $A \subset \mathbb{R}^n$.

Definition 5.12.

$$\overline{\dim}_B A := \limsup_{\delta \to 0} \frac{\log N_\delta^B(A)}{|\log \delta|},$$

$$\underline{\dim}_B A := \liminf_{\delta \to 0} \frac{\log N_\delta^B(A)}{|\log \delta|}$$

are called *upper* and *lower box dimension* of A, respectively. If the limits agree, then the common value $\dim_B A$ is the *box dimension* of A.

Theorem 5.13. *For any bounded $A \subset \mathbb{R}^n$ we have*

$$\overline{\dim}_B A = \overline{\dim}_E A = \overline{\dim}_\varrho A = \overline{\dim}_M A$$

$$\underline{\dim}_B A = \underline{\dim}_E A = \underline{\dim}_\varrho A = \underline{\dim}_M A.$$

The proof of the first equalities is similar to the above proofs for the others, when comparing intersecting boxes with outer covering balls and covering balls with the collection of neighboring cubes of the same size.

This theorem shows, in particular, that the box dimension does not depend on the choice of the coordinate system for the lattices.

The method of box counting is widely used in statistical procedures for estimating dimensions. (For Hausdorff dimensions, in general, it does not work well.)

5.4 Comparison with Hausdorff and packing dimensions

Let (X, ϱ) be a metric space and $A \subset X$. From 2.34 and 5.1 we know the relationships

$$\dim_H A \leq \dim_P A \quad \text{and} \quad \dim_H A \leq \underline{\dim_E} A.$$

Theorem 5.14.

$$\dim_P A \leq \overline{\dim}_\varrho A.$$

Proof. Let $0 \leq t < s < \dim_P A$. Then we have $\widetilde{\mathcal{P}}^s(A) = \infty$, i.e., $\widetilde{\mathcal{P}}_0^s(A) = \infty$. Thus, for any sufficiently small $\delta > 0$, we can find a 2δ-ball packing $(B(x_i, r_i))_i$ of A such that

$$\sum_i |2r_i|^s > 1.$$

Let $\nu_k \leq \infty$ be the number of balls with $2^{-(k+1)} < 2r_i \leq 2^{-k}$. Then we get

$$\sum_k \nu_k 2^{-ks} > 1,$$

and there must be an $l \in \mathbb{N}$ such that $\nu_l > 2^{lt}(1 - 2^{t-s})$. (Otherwise, $\sum_k \nu_k 2^{-ks} \leq \sum_k 2^{k(t-s)}(1 - 2^{t-s}) = 2^{t-s} < 1$, which contradicts the above inequality.)

The ν_l associated balls all contain the ball with the same center and radius $2^{-(l+2)}$. Hence,

$$N_{2^{-(l+2)}}(A) \geq \nu_l > 2^{lt}(1 - 2^{t-s})$$

and consequently,

$$\frac{\log N_{2^{-(l+2)}}(A)}{|\log 2^{-(l+2)}|} > \frac{l\, t \log 2 + \log(1 - 2^{t-s})}{(l+2) \log 2}.$$

By construction, for any $\delta > 0$ there exists an $l \in \mathbb{N}$ satisfying the last inequality, and these numbers l increase as $\delta \to 0$. From this we infer

$$\overline{\dim}_\varrho A = \limsup_{\delta \to 0} \frac{\log N_\delta(A)}{|\log \delta|} \geq \limsup_{l \to \infty} \frac{\log N_{2^{-(l+2)}}(A)}{|\log 2^{-(l+2)}|} \geq t$$

for any $t < \dim_P A$, i.e.,

$$\overline{\dim}_\varrho A \geq \dim_P A.$$

$\square$

If we take a modification of the upper metric dimension, then it agrees with the packing dimension. This provides a *measure free representation of packing dimension*:

Theorem 5.15.

$$\dim_P A = \inf \left\{ \sup_i \overline{\dim}_\varrho A_i : A = \bigcup_{i=1}^\infty A_i, \ A_i \text{ totally bounded} \right\},$$

for any subset A of a metric space (X, ϱ).

Proof. If $\dim_P A < s$, then $\widetilde{\mathcal{P}}^s(A) = 0$ and there exists a representation $A = \bigcup_i A_i$ such that $\sum_{i=1}^\infty \widetilde{\mathcal{P}}_0^s(A_i) < \infty$. This implies $\widetilde{\mathcal{P}}_0^s(A_i) < \infty$ and thus $\widetilde{\mathcal{P}}_{2\delta}^s(A_i) < \infty$, $\delta < \delta_i$, for any i. By construction of $\widetilde{\mathcal{P}}_\delta^s$,

$$(2\delta)^s N_\delta(A_i) \leq \widetilde{\mathcal{P}}_{2\delta}^s(A_i) < \infty, \quad \delta < \delta_i.$$

Therefore, the sets A_i are totally bounded and

$$\limsup_{\delta \to 0} \delta^s N_\delta(A_i) \leq \widetilde{\mathcal{P}}_0^s(A_i) < \infty.$$

The last inequality implies

$$\limsup_{\delta \to 0} |\log \delta| \left(\frac{\log N_\delta(A_i)}{|\log \delta|} - s \right) \leq \log \widetilde{\mathcal{P}}_0^s(A_i) < +\infty,$$

which can only be true for

$$\limsup_{\delta \to 0} \frac{\log N_\delta(A_i)}{|\log \delta|} \leq s.$$

Since the last inequality holds for every i, we obtain

$$\sup_i \{\overline{\dim}_\varrho A_i\} \leq s.$$

Moreover, for any $s > \dim_P A$ we have got a sequence A_i of totally bounded sets with the above properties. Hence,

$$\inf \left\{ \sup_i \overline{\dim}_\varrho A_i : A = \bigcup_{i=1}^\infty A_i, \ A_i \text{ totally bounded} \right\} \leq \dim_P A.$$

In order to prove the opposite inequality we first assume that a representation $A = \bigcup_i A_i$, A_i totally bounded, is impossible. Then the infimum equals $+\infty \geq \dim_P A$. Otherwise for any such representation we get

$$\dim_P A \leq \sup_i \dim_P A_i \leq \sup_i \overline{\dim}_\varrho A_i$$

using 2.35 (ii) and 5.14. Consequently,

$$\dim_P A \leq \inf \left\{ \sup_i \{\overline{\dim}_\varrho A_i\} : A = \bigcup_{i=1}^\infty A_i \,, \; A_i \text{ totally bounded} \right\}.$$

$\square$

Remark 5.16. According to Theorem 5.9, on the right-hand side of the statement $\overline{\dim}_\varrho A_i$ may be replaced by $\overline{\dim}_E A_i$. Under some additional conditions a modification of the upper entropy dimension is not needed:

Theorem 5.17. *If A is a closed subset of a complete metric space (X, ϱ) such that $\overline{\dim}_E(A \cap G) = \overline{\dim}_E A$ for any open set G intersecting A, then we have*

$$\dim_P A = \overline{\dim}_E A.$$

Proof. We will use the Baire category theorem from topology: A complete metric space cannot be represented as countable union of nowhere dense sets. Thus, if $A = \bigcup_i F_i$ is a representation with closed sets F_i, then for some $k \in \mathbb{N}$ there exists an open set G with $A \cap G \subset F_k$.

Furthermore,

$$\overline{\dim}_E A \geq \overline{\dim}_E F_k \geq \overline{\dim}_E(A \cap G) = \overline{\dim}_E A$$

by assumption. Hence,

$$\overline{\dim}_E F_k = \overline{\dim}_E A.$$

According to Theorem 5.15,

$$\dim_P A = \inf \left\{ \sup_i \{\overline{\dim}_\varrho A_i\} : A = \bigcup_{i=1}^\infty A_i \,, \; A_i \text{ totally bounded} \right\}.$$

It is not difficult to see that for any totally bounded $B \subset X$ we have $\overline{\dim}_\varrho B = \overline{\dim}_\varrho \overline{B}$. Further, since A is closed, the representation $A = \bigcup_i A_i$ implies $A = \bigcup_i \overline{A}_i$, and $\overline{\dim}_\varrho \overline{A}_i = \overline{\dim}_E A_i$ for totally bounded A_i. Setting $\overline{A}_i =: F_i$ we obtain

$$\dim_P A = \inf \left\{ \sup_i \overline{\dim}_E F_i : A = \bigcup_i F_i \,, \; F_i \text{ closed and totally bounded} \right\}$$

$$\geq \overline{\dim}_E A,$$

according to the above equality for $\overline{\dim}_E F_k$.

On the other hand, in view of Theorem 5.14,

$$\dim_P A \leq \overline{\dim}_\varrho A = \overline{\dim}_E A.$$

$\square$

In the next section we will provide examples for

$$\underline{\dim}_E A < \dim_P A \quad \text{and for} \quad \dim_P A < \underline{\dim}_E A.$$

5.5 Examples

Example 5.18 (Cantor sets). From 2.37 we know that the Cantor set $C(a)$, with contraction ratio $0 < a < 1/2$, possesses the Hausdorff dimension $\dim_H C(a) = \log 2/|\log a|$.

Moreover, since $C(a)$ may be covered by 2^k closed intervals of length a^k, we obtain

$$\overline{\dim}_E C(a) = \limsup_{\delta \to 0} \frac{\log n_\delta(C(a))}{|\log \delta|} \leq \limsup_{k \to \infty} \frac{\log 2^k}{|\log a^k|} = \frac{\log 2}{|\log a|},$$

i.e.,

$$\boxed{\dim_H C(a) = \dim_P C(a) = \dim_E C(a) = \frac{\log 2}{|\log a|}.}$$

Example 5.19 (Harmonic sequence). Set

$$A := \left\{ 0, 1, \frac{1}{2}, \frac{1}{3}, \ldots \right\}.$$

Then we get

$$\boxed{\dim_H A = \dim_P A = 0}$$

(and $\mathcal{H}^0(A) = \infty$). We will show now that

$$\boxed{\dim_E A = \tfrac{1}{2}.}$$

If $\frac{1}{k(k+1)} < 2\delta \leq \frac{1}{(k-1)k}$ then at least k intervals of length 2δ are needed in order to cover A. Hence, $n_\delta(A) \geq k$ and therefore

$$\underline{\dim}_E A = \liminf_{\delta \to 0} \frac{\log n_\delta(A)}{|\log \delta|} \geq \liminf_{k \to \infty} \frac{\log k}{\log(k(k+1))} = \frac{1}{2}.$$

Further, $[0, 1/k]$ may be covered by $k + 1$ intervals of length 2δ, and the points $\{\frac{1}{k-1}, \frac{1}{k}, \ldots, 1\}$ by $k - 1$ intervals of length 2δ. Hence, $n_\delta(A) \leq k + 1 + k - 1$ and

$$\overline{\dim}_E A = \limsup_{\delta \to 0} \frac{\log n_\delta(A)}{|\log \delta|} \leq \lim_{k \to \infty} \frac{\log 2k}{\log k(k-1)} = \frac{1}{2}.$$

Example 5.20 (Cantor sets with varying contraction ratios). In each construction step n of a generalized Cantor set C the contraction ratio a_n of the two linear mapping is now chosen possibly different. We suppose that $0 < a_n < 1/2$ and

$$\underline{a} := \varliminf_{n \to \infty} a_n < \varlimsup_{n \to \infty} a_n =: \overline{a}.$$

Then the length of each of the 2^n intervals in the n-step equals $a_1 \cdots a_n$. If

$$a_1 \cdots a_{n+1} \leq 2\delta < a_1 \cdots a_n,$$

then the covering number $n_\delta(C)$ satisfies

$$2^n < n_\delta(C) \leq 2^{n+1},$$

i.e.,

$$\frac{n \log 2}{\sum\limits_{k=1}^{n+1} |\log a_k| - \log 2} < \frac{\log n_\delta(C)}{|\log \delta|} \leq \frac{(n+1) \log 2}{\sum\limits_{k=1}^{n} |\log a_k| - \log 2}.$$

Therefore we infer

$$\varliminf_{\delta \to 0} \frac{\log n_\delta(C)}{|\log \delta|} = \varliminf_{n \to \infty} \frac{n \log 2}{\sum\limits_{k=1}^{n} |\log a_k|} = \frac{\log 2}{\varliminf\limits_{n \to \infty} \frac{1}{n} \sum\limits_{k=1}^{n} |\log a_k|} =: \underline{d}$$

and similarly,

$$\varlimsup_{\delta \to 0} \frac{\log n_\delta(C)}{|\log \delta|} = \frac{\log 2}{\varlimsup\limits_{n \to \infty} \frac{1}{n} \sum\limits_{k=1}^{n} |\log a_k|} =: \overline{d}.$$

Note that

$$\frac{\log 2}{|\log \underline{a}|} \leq \underline{d} \leq \overline{d} \leq \frac{\log 2}{|\log \overline{a}|}$$

and any admissible pair $\underline{d}, \overline{d}$ may be constructed by a suitable choice of the contraction ratios a_n. We have got

$$\boxed{\underline{\dim}_E C = \underline{d}} \quad \text{and} \quad \boxed{\overline{\dim}_E C = \overline{d}}.$$

Next we will show that

$$\boxed{\dim_P C = \overline{\dim}_E C}$$

using Theorem 5.17. If $G \subset [0,1]$ is an open interval intersection C then for sufficiently large m it contains one of the intervals I_m in the mth step. Similarly as above one shows

$$\overline{\dim}_E(C \cap I_m) = \overline{d}.$$

(Instead of 2^n covering intervals one chooses only 2^{n-m} intervals of the same size, and the upper limits coincide.) Therefore,

$$\overline{\dim}_E(C \cap G) = \overline{d} = \overline{\dim}_E C$$

for any open G in $[0, 1]$, so that Theorems 5.17 may be applied.

Finally, we will prove

$$\boxed{\dim_H C = \underline{d}}$$

by means of the mass distribution principle 3.8. ($\dim_H C \leq \underline{\dim}_E C = \underline{d}$ is already known.) Let $0 < s < \underline{d}$ be arbitrary. It suffices to construct a finite Borel measure μ on C such that $\overline{\mathcal{D}}^s \mu(x) = 0$, $x \in C$. We apply the outer measure construction 1.12 on $[0, 1]$ choosing as covering system $\mathcal{C}$ the set of all closed intervals I^n together the open cutout intervals $\widetilde{I}^n$ arising in the n-th construction steps, $n = 0, 1, \ldots$ For such intervals we set

$$\tau(I^n) := 2^{-n}, \quad \tau(\widetilde{I}^n) := 0.$$

It is easy to see that the corresponding outer measure μ is metric, i.e., μ is a Borel measure on $I_0 = [0, 1]$. By construction,

$$\mu(I_n) = \tau(I_n).$$

(This follows from the fact that $\tau(I_n) = \sum_{i=1}^{\infty} \tau(C_i)$ if $I^n = \bigcup_{i=1}^{\infty} C_i$ for some disjoint $C_i \in \mathcal{C}$.)

Since μ has no mass on the cutouts $\widetilde{I}_n$ at each step, the support of μ agrees with the generalized Cantor set C and $\mu(C) = \mu(I_0) = 1$. In order to calculate the densities $\overline{\mathcal{D}}^s \mu(x)$, for given $r > 0$ define n by

$$a_1 \cdots a_{n+1} \leq 2r < a_1 \cdots a_n.$$

Then for any $x \in C$ the "ball" $B(x, r)$ intersects at most one of the intervals I_n in the n-th step. Taking into regard that μ has no mass between these intervals, we obtain

$$\frac{\mu(B(x, r))}{r^s} \leq \frac{\mu(I_n)}{r^s} \leq \frac{2^{-n} 2^s}{a_1^s \cdots a_{n+1}^s}$$

$$= \exp\left\{-n \ln 2 + s \ln 2 + s \sum_{k=1}^{n+1} |\ln a_i|\right\}$$

$$= \exp\left\{n\left(-\ln 2 + \frac{1}{n} s \ln 2 + s\frac{1}{n} \sum_{k=1}^{n+1} |\ln a_i|\right)\right\}.$$

Using that the upper limit as $n \to \infty$ of the second factor in the exponent is equal to

$$-\ln 2 + s\,\underline{d}^{-1} \ln 2 < 0$$

in view of the above definition of $\underline{d}$, we infer

$$\overline{\mathcal{D}}^s \mu(x) = \varlimsup_{r \to 0} \frac{\mu(B(x,r))}{(2r)^s} \leq \varlimsup_{n \to \infty} \exp\{\dots\}$$
$$= \exp\left\{\varlimsup_{n \to \infty} \dots\right\} = \exp\{-\infty\} = 0$$

for any $x \in C$. Consequently, $\dim_H C \geq s$ for any $s < \underline{d}$.

Thus, in case $\underline{d} < \overline{d}$ we have got

$$\dim_H C = \underline{\dim}_E C < \overline{\dim}_E C = \dim_P C.$$

Remark 5.21. Examples 5.5.2 and 5.5.3 demonstrate that both inequalities $\underline{\dim}_E A > \dim_P A$ or $\underline{\dim}_E A < \dim_P A$ are possible.

Chapter 6

Dimensions of Borel Measures

6.1 Pointwise dimensions

In Chapter 3 we have seen that Hausdorff and packing dimensions of Borel sets are closely related to upper and lower densities of mass distributions on these sets.

Now locally finite Borel measures μ in the metric space (X, d) will be the primary subject of study, i.e., where every point has a neighborhood with finite measure. We are interested in their dimension properties. The link with Hausdorff and packing dimensions of underlying sets will be established using the density results.

We start with *local dimension characteristics* of μ.

Definition 6.1.
$$\underline{\dim}\mu(x) := \liminf_{r \to 0} \frac{\log \mu(B(x, r))}{\log r},$$
$$\overline{\dim}\mu(x) := \limsup_{r \to 0} \frac{\log \mu(B(x, r))}{\log r}$$
are called *lower* and *upper pointwise dimension* of μ at $x \in X$, respectively.

Obviously, $\underline{\dim}\mu(x) \leq \overline{\dim}\mu(x)$. Both these notions may be expressed in terms of densities:

Exercise 6.1.

(i) $\underline{\dim}\mu(x) = \sup\{s \geq 0 : \overline{\mathcal{D}}^s\mu(x) = 0\} = \inf\{s \geq 0 : \overline{\mathcal{D}}^s\mu(x) = \infty\}$,

(ii) $\overline{\dim}\mu(x) = \sup\{s \geq 0 : \underline{\mathcal{D}}^s\mu(x) = 0\} = \inf\{s \geq 0 : \underline{\mathcal{D}}^s\mu(x) = \infty\}$.

(This can be proved similarly as Proposition 5.6)

Definition 6.2. The measure μ is said to be *dimension regular*, if $\underline{\dim}\mu(x) = \overline{\dim}\mu(x)$ at μ-a.a. x.

In Chapter 7 we will provide examples of such measures.

6.2 Global dimensions

Taking essential extrema of the above local dimensions we obtain the corresponding global versions: Let μ again be a locally finite Borel measure in (X, d).

Definition 6.3.

(i) $\dim_H \mu := \text{ess sup}_{x \in X} \underline{\dim}\mu(x)$
$$= \inf\{s \geq 0 : \underline{\dim}\mu(x) \leq s \text{ for } \mu\text{-a.a. } x \in X\}$$
is called *Hausdorff dimension* of the Borel measure μ.

$\underline{\dim}_H \mu := \text{ess inf}_{x \in X} \underline{\dim}\mu(x)$
$$= \sup\{s \geq 0 : \underline{\dim}\mu(x) \geq s \text{ for } \mu\text{-a.a. } x \in X\}$$
is the *lower Hausdorff dimension* of μ.

(ii) $\dim_P \mu := \text{ess sup}_{x \in X} \overline{\dim}\mu(x)$
is called *packing dimension* of μ.

$\underline{\dim}_P \mu := \text{ess inf}_{x \in X} \overline{\dim}\mu(x)$
is its *lower packing dimension*.

For dimension regular μ we get

$$\dim_H \mu = \dim_P \mu \quad \text{and} \quad \underline{\dim}_H \mu = \underline{\dim}_P \mu. \tag{6.1}$$

The Density theorems 3.7 and 3.10 provide the following equivalent expressions for (i) and (ii) in terms of Hausdorff and packing dimensions of underlying sets.

Theorem 6.4. *Let μ be a Borel measure on the metric space (X, d) with $0 < \mu(X) < \infty$. Then we have:*

(i) $\underline{\dim}_H \mu = \inf\{\dim_H B : \mu(B) > 0, \ B \in \mathfrak{B}(X)\}$,

(ii) $\dim_H \mu = \inf\{\dim_H B : \mu(X \setminus B) = 0, \ B \in \mathfrak{B}(X)\}$,

(iii) $\underline{\dim}_P \mu = \inf\{\dim_P B : \mu(B) > 0, \ B \in \mathfrak{B}(X)\}$, ,

(iv) $\dim_P \mu = \inf\{\dim_P B : \mu(X \setminus B) = 0, \ B \in \mathfrak{B}(X)\}$.

Proof. (i) Denote $\underline{\dim}_H \mu =: \alpha$. For $\beta < \alpha$ set

$$A_\beta := \{x \in X : \underline{\dim}\mu(x) > \beta\}.$$

By means of Proposition 3.5 it can be shown that A_β is a Borel set. Further, $\mu(X \setminus A_\beta) = 0$, since otherwise α would not be the essential infimum.

Choose any $B \in \mathfrak{B}(X)$ with $\mu(B) > 0$ and define the Borel set $C_\beta := A_\beta \cap B$. Obviously,

$$\dim_H B \geq \dim_H C_\beta.$$

Since $\mu(C_\beta) = \mu(B)$, the measure μ may be interpreted as a mass distribution on C_β. Moreover, in view of 6.1 and $\beta < \alpha$ we have $\overline{D}^\beta \mu(x) = 0$ for any $x \in C_\beta$. Thus, the mass distribution principle (see Theorem 3.7 (i) together with Remark 3.8) implies $\dim_H C_\beta \geq \beta$. Consequently,

$$\dim_H B \geq \beta \quad \text{for any } \beta < \alpha,$$

i.e.,

$$\dim_H B \geq \alpha \quad \text{for any } B \in \mathfrak{B}(X) \text{ with } \mu(B) > 0,$$

and therefore

$$\inf\{\dim_H B : B \in \mathfrak{B}(X), \ \mu(B) > 0, \} \geq \alpha.$$

In order to prove the opposite inequality we now choose $\beta > \alpha$ and the Borel set

$$B_\beta := \{x \in X : \underline{\dim}\mu(x) < \beta\}.$$

Since α is the essential infimum and $\alpha < \beta$ we get $\mu(B_\beta) > 0$. 6.1 implies $\overline{D}^\beta \mu(x) = \infty$, for any $x \in B_\beta$. Thus, we can apply the mass distribution principle 3.7 (ii) to μ and B_β in order to obtain $\dim_H B_\beta \leq \beta$. It follows that

$$\inf\{\dim_H B : B \in \mathfrak{B}(X), \ \mu(B) > 0\} \leq \beta,$$

since B_β is one of the sets B. Letting $\beta \downarrow \alpha$ we infer the desired inequality, i.e., assertion (i) is proved.

(ii) Denote now $\dim_H \mu =: \alpha$. First choose $\beta > \alpha$ and B_β as above. In this case we get $\mu(X \setminus B_\beta) = 0$, since otherwise α would not be the essential supremum. Further, $\dim_H B_\beta \leq \beta$ by the above arguments, because $\beta > \underline{\dim}_H \mu$. Hence, we obtain

$$\inf\{\dim_H B : B \in \mathfrak{B}(X), \ \mu(X \setminus B) = 0\} \leq \beta$$

using that B_β is one of the sets B. Letting $\beta \downarrow \alpha$ we conclude that the left-hand side does not exceed α.

For proving the opposite inequality we now consider $\beta < \alpha$ and the set $A_\beta := \{x \in X : \underline{\dim}\mu(x) > \beta\}$ as above. Here we get $\mu(A_\beta) > 0$, since otherwise α would not be the essential supremum. For $B \in \mathfrak{B}(X)$ such that $\mu(X \setminus B) = 0$ and the Borel set $C_\beta := A_\beta \cap B$ we obtain $\mu(C_\beta) > 0$

and by the first part of the proof of (i), $\dim_H C_\beta \geq \beta$, i.e., $\dim_H B \geq \beta$. Since $\beta > \alpha$ was arbitrary, it follows that $\dim_H B \geq \alpha$ for any $B \in \mathfrak{B}(X)$ with $\mu(X \setminus B) = 0$.

Therefore the infimum of $\dim_H B$ over such B is not less than α, which completes the proof of (ii).

The arguments for (iii) and (iv) are the same, when replacing upper densities and Theorem 3.7 by lower densities and Theorem 3.10. $\qquad\square$

In the literature the equalities in Theorem 6.4 are often used as definitions for Hausdorff and packing dimensions of finite Borel measures.

Definition 6.5. If

$$\underline{\dim}_{(\cdot)}\mu = \dim_{(\cdot)}$$

for one of the subscriptions H or P, then μ is said to be of *exact Hausdorff or packing dimension*, respectively.

Note that the inequality $\underline{\dim}_{(\cdot)} \leq \dim_{(\cdot)}$ is always true.

If μ is dimension regular and of exact dimension then the local dimensions at μ-a.a. points agree with the global dimensions, say D, of μ. But even in this case, in general, the set

$$A_\alpha := \{x \in X : \lim_{r \to 0} \frac{\log \mu(B(x,r))}{\log r} = \alpha\}$$

for certain $\alpha \neq D$ is not empty. The family $\{\dim_H A_\alpha\}$ or $\{\dim_P A_\alpha\}$ over these α is called *multifractal spectrum* and is of great physical interest. For various models in fractal geometry it is determined by means of suitable mass distribution constructed on the sets A_α. For more details see, e.g., [Falconer (1997)] and the references therein, or the survey [Olsen (2000)].

Recall that $\dim_H \mu = \inf\{\dim_H B : B \in \mathfrak{B}(X), \mu(X \setminus B) = 0\}$ and $\dim_P \mu = \inf\{\dim_P B : B \in \mathfrak{B}(X), \mu(X \setminus B) = 0\}$. If X is separable, then the support $\operatorname{supp}\mu$ of μ may be interpreted as one of the sets B. This leads to the following.

Corollary 6.6.

$$\dim_H \mu \leq \dim_H \operatorname{supp}\mu \quad and \quad \dim_P \mu \leq \dim_P \operatorname{supp}\mu$$

for any finite Borel measure μ on a separable metric space (X, d).

Examples for the strict inequality will be provided in Chapter 7.

6.3 A variational principle for Hausdorff and packing dimensions

We now will prove a certain counterpart to Theorem 6.4 where the dimensions of a measure were expressed in terms of the infimum of dimensions of certain underlying Borel sets. Conversely, the dimensions of a Borel set may be calculated by means of the supremum of dimensions of Borel measures concentrated on the set. Here we use the notation $\mathcal{M}(A)$ for the *family of finite Borel measures on (X,d) with support in A*.

Theorem 6.7. *Let (X,d) be a Polish space and $A \in \mathfrak{B}(X)$. Then we have the following:*

$$\text{(i)} \quad \dim_H A = \sup_{\mu \in \mathcal{M}(A)} \dim_H \mu,$$

$$\text{(ii)} \quad \dim_P A = \sup_{\mu \in \mathcal{M}(A)} \dim_P \mu.$$

Proof. (i) For any $\mu \in \mathcal{M}(A)$ we have $\mu(X \setminus A) = 0$, hence,

$$\dim_H \mu = \inf\{\dim_H B : \mu(X \setminus B) = 0,\ B \in \mathfrak{B}(X)\} \le \dim_H A.$$

This implies

$$\sup_{\mu \in \mathcal{M}(A)} \dim_H \mu \le \dim_H A.$$

If $\dim_H A = 0$, then we are done.

Suppose that $\dim_H A > 0$. In order to prove the opposite inequality we choose an arbitrary $0 < t < s := \dim_H A$, for which $\mathcal{H}^t(A) = \infty$. Then by Theorem 2.40 there exists a compact set $K \subset A$ such that $0 < \mathcal{H}^t(K) < \infty$. The measure

$$\mu_t := \mathcal{H}^t \llcorner K$$

is an element of $\mathcal{M}(A)$. The Density theorem 3.9 for Hausdorff measures implies $2^{-t} \le \mathcal{D}^t \mu_t(x) \le 1$ at μ_t-a.a. x. Therefore in view of 6.1 we get

$$\dim_H \mu_t = t.$$

Consequently,

$$\sup_{\mu \in \mathcal{M}(A)} \dim_H \mu \ge \dim_H \mu_t = t.$$

Letting $t \uparrow s$ we obtain the desired inequality.

The proof of (ii) is analogous, when replacing Hausdorff measures by packing measures and using the corresponding part of Theorem 2.40. $\qquad\square$

In the next chapters we will enlighten these notions and relationships by special fractal models. We start with so-called self-similar sets and measures and certain extensions which have been studied in great detail over many years.

Chapter 7

Attractors of Iterated Function Systems

7.1 Generating principle in boundedly compact spaces

In the preceding sections we have repeatedly used linear Cantor-type sets in order to demonstrate some dimension relationships. Their generating principle may be easily extended to higher-dimensional sets: instead of the primary unit interval and the mappings S_1, S_2 one takes other geometric figures and finitely many similarities $S_1, \dots, S_N$. We illustrate a few classical cases in $\mathbb{R}^2$ in form of Figures 7.1–7.5 below. The pictures describe some first construction steps with appropriate primary sets. Their continuations approximate the self-similar limit sets for the corresponding similarities S_i. The general construction model behind will now be introduced. (See

Fig. 7.1   Cantor Set $C = S_1(C) \cup S_2(C)$

Theorem 7.5 below.)

Throughout this chapter (X, d) is a *complete metric space*. Sometimes we will specify to *boundedly compact spaces* (where all bounded closed sets are compact).

We use the notions of Lipschitz mappings $f : X \to X$ or of similarities $S : X \to X$ (recall Definition 2.6). The *Lipschitz constant* of f is given by

$$\operatorname{Lip}(f) := \sup_{x \neq y} \frac{d(f(x), f(y))}{d(x, y)}.$$

Fig. 7.2   Sierpinski gasket $SG = S_1(SG) \cup S_2(SG) \cup S_3(SG)$

Note that similarities S with ratio r are special Lipschitz mappings and $\mathrm{Lip}(S) = r$. A mapping $f : X \to X$ is called a *contraction* if $\mathrm{Lip}(f) < 1$.

The *space of non-empty compact subsets of* X is denoted by $\mathcal{K}$.

Definition 7.1. Let $\varphi_1, \ldots, \varphi_N$ be contractions of X, $N \geq 2$. We define a map $\varphi : \mathcal{K} \to \mathcal{K}$ as follows:

$$\varphi(K) := \varphi_1(K) \cup \ldots \cup \varphi_N(K), \quad K \in \mathcal{K}.$$

The map φ may be considered as a contraction on $\mathcal{K}$, if this space is provided with a suitable metric.

Definition 7.2. If K and L are compact subsets of the space (X, d), the set-function

$$d_H(K, L) := \max\{\max_{x \in L} d(x, K), \max_{y \in K} d(y, L)\}$$

is called *Hausdorff distance* between K and L.

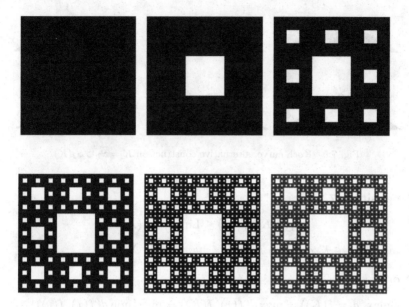

Fig. 7.3   Sierpinski carpet $SC = S_1(SC) \cup S_2(SC) \cup \ldots \cup S_8(SC)$

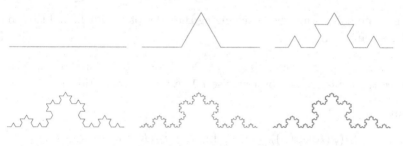

Fig. 7.4   Koch curve $K = S_1(K) \cup S_2(K) \cup S_3(K) \cup S_4(K)$

Exercise 7.1.

(i) d_H is a metric on $\mathcal{K}$.

(ii) $d_H(\{x\}, \{y\}) = d(x, y)$, for $x, y \in X$.

(iii) $d_H(K, L) = \min\{\varepsilon \geq 0 : K \subset L_\varepsilon,\ L \subset K_\varepsilon\}$, for the ε-parallel sets of K and L.

(iv) $d_H\left(\bigcup_{i=1}^{N} K_i, \bigcup_{i=1}^{N} L_i \right) \leq \max_i d_H(K_i, L_i)$.

The Hausdorff distance has the following behavior under Lipschitz mappings.

Fig. 7.5   Koch curve, alternative construction $K = \widetilde{S}_1 \cup \widetilde{S}_2(K)$

Proposition 7.3.

$$d_H(f(K), f(L)) \leq \mathrm{Lip}(f) \, d_H(K, L)$$

for any $K, L \in \mathcal{K}$ and any Lipschitz mapping $f : X \to X$.

Proof.

$$\max_{z \in f(L)} d(z, f(K)) = \max_{x \in L} d(f(x), f(K)) = \max_{x \in L} \{ \min_{y \in K} d(f(x), f(y)) \}$$

$$\leq \max_{x \in L} \{ \mathrm{Lip}(f) \min_{y \in K} d(x, y) \} = \mathrm{Lip}(f) \max_{x \in L} d(x, K) \,.$$

The same holds true when exchanging the roles of K and L, and this completes the proof.   □

Let now $\varphi_1, \dots \varphi_n$ and φ be as in Definition 7.1. Then the *contraction property* of φ is implied by Exercise 7.1 (iv) and Proposition 7.3:

Corollary 7.4.

$$d_H(\varphi(K), \varphi(L)) \leq \max_{1 \leq i \leq N} \mathrm{Lip}(\varphi_i) \, d_H(K, L) \,, \quad K, L \in \mathcal{K} \,.$$

In the sequel we will use the notations

$$r_i := \mathrm{Lip}(\varphi_i) \ i = 1, \dots N \,, \quad r_{max} := \max_i r_i \,, \quad r_{min} := \min_i r_i \,.$$

One can show that the metric space $(\mathcal{K}, d_H)$ is complete if (X, d) is complete. Then Banach's fixed point theorem implies the existence of a unique element $A \in \mathcal{K}$, which is invariant under the mapping φ. We will not give the completeness proof here, since the fixed point A will be determined in a more direct way. We start with the more special case of boundedly compact spaces. In the next section extensions to arbitrary complete metric spaces will be treated.

Theorem 7.5. *Let $\varphi_1, \ldots, \varphi_N$, $N \geq 2$, be contractions of a boundedly compact metric space (X, d) and let $\varphi(K) := \varphi_1(K) \cup \ldots \cup \varphi_N(K)$, $K \in \mathcal{K}$. Then there is a unique non-empty compact set A such that*

$$\varphi(A) = A.$$

Moreover, this fixed point may be determined by

$$A = \bigcap_{n=1}^{\infty} \varphi^n(K)$$

for any $K \in \mathcal{K}$ such that $\varphi(K) \subset K$.

Proof. We will first show that for any $z \in X$ and sufficiently large $R > 0$, the ball $B(z, R)$ has the property $\varphi(B(z, R)) \subset B(z, R)$, i.e., compact K as in the second assertion do exist. Let x_i be the fixed point of the mapping φ_i, $i = 1, \ldots, N$, and $d_{max} := \max_i d(x_i, z)$. If $x \in B(z, R)$, then

$$
\begin{aligned}
d(\varphi_i(x), z) &\leq d(\varphi_i(x), \varphi_i(x_i)) + d(\varphi_i(x_i), z) \\
&\leq r_{max} \, d(x, x_i) + d(x_i, z) \\
&\leq r_{max} \, d(x, x_i) + d_{max} \\
&\leq r_{max}(d(x, z) + d(z, x_i)) + d_{max} \\
&\leq r_{max}(R + d_{max}) + d_{max} \\
&= r_{max} R + (r_{max} + 1)d_{max} < R,
\end{aligned}
$$

provided $R > (r_{max} + 1)d_{max}/(1 - r_{max})$, which yields the above assertion.

Now choose any $K \in \mathcal{K}$ with $\varphi(K) \subset K$. This inclusion implies $\varphi^{n+1}(K) \subset \varphi^n(K)$, $n \in \mathbb{N}$, and these sets are all compact. Hence,

$$A := \bigcap_{n=1}^{\infty} \varphi^n(K)$$

is also a non-empty compact set and satisfies $\varphi(A) = A$.

It remains to prove uniqueness of the fixed point. Suppose $A' \in \mathcal{K}$ and $\varphi(A') = A'$. Then we get

$$d_H(A, A') = d_H(\varphi(A), \varphi(A')) \leq r_{max} \, d_H(A, A'),$$

and therefore $d_H(A, A') = 0$, since $r_{max} < 1$. $\qquad\square$

The set A may be generated by iterated applications of the function system $\varphi_1, \ldots \varphi_N$ to any starting compact set K_0:

Proposition 7.6.

$$\lim_{n \to \infty} d_H(\varphi^n(K_0), A) = 0, \quad K_0 \in \mathcal{K}.$$

Proof. $d_H(\varphi^n(K_0), A) = d_H(\varphi^n(K_0), \varphi^n(A)) \leq r_{max}^n \, d_H(K_0, A)$, which vanishes at infinity. $\qquad\qquad\qquad\qquad\qquad\qquad\qquad\qquad\qquad\qquad$ $\square$

Therefore A is also called *attractor of the iterated function system* (*briefly IFS*) $\varphi_1, \ldots, \varphi_N$. If all mappings are similarities, then A is called *self-similar set*.

Proposition 7.6 provides the *first method for simulating* A: Take $K_0 := \{x_0\}$ for an arbitrary starting point $x_0 \in X$ and use

$$A = \lim_{n\to\infty} \varphi^n(\{x_0\})$$

for computer approximations.

7.2   An encoding procedure and extension to arbitrary complete metric spaces

The point structure of the above attractor A may be described by means of an appropriate coding space which will be introduced below. We first take an arbitrary $K \in \mathcal{K}$ with $\varphi(K) \subset K$ and decompose

$$A = \bigcap_{i=1}^{\infty} \varphi^n(K) = \bigcap_{i=1}^{\infty} \bigcup_{i_1,\ldots,i_n\in\{1,\ldots,N\}} \varphi_{i_1} \circ \cdots \circ \varphi_{i_n}(K)$$

$$= \bigcup_{(i_1,i_2,\ldots)\in\{1,\ldots,N\}^n} \bigcap_{i=1}^{\infty} \varphi_{i_1} \circ \cdots \circ \varphi_{i_n}(K).$$

By the choice of K we have

$$\varphi_{i_1} \circ \cdots \circ \varphi_{i_n}(K) \subset \varphi_{i_1} \circ \cdots \circ \varphi_{i_{n-1}}(K).$$

These sets are compact and their diameters tend to zero as $n \to \infty$. Therefore the set $\bigcap_{n=1}^{\infty} \varphi_{i_1} \circ \cdots \circ \varphi_{i_n}(K)$ consists of a single point denoted by $x_{i_1,i_2,\ldots}$. Moreover,

$$d_H(\{x_{i_1,i_2,\ldots}\}, \{\varphi_{i_1} \circ \cdots \circ \varphi_{i_n}(x_0)\}) \leq d_H(\{x_{i_1,i_2,\ldots}\}, \varphi_{i_1} \circ \cdots \circ \varphi_{i_n}(K))$$

$$+ d_H(\varphi_{i_1} \circ \cdots \circ \varphi_{i_n}(K), \{\varphi_{i_1} \circ \cdots \circ \varphi_{i_n}(x_0)\}),$$

which tends to 0 as $n \to \infty$. Hence,

$$\lim_{n\to\infty} \varphi_{i_1} \circ \cdots \circ \varphi_{i_n}(x_0) = x_{i_1,i_2,\ldots}$$

for any starting point $x_0 \in X$, and we infer the following.

Proposition 7.7.

$$A = \bigcup_{i_1,\ldots,i_n\in\{1,\ldots,N\}^n} \lim_{n\to\infty} \varphi_{i_1} \circ \cdots \circ \varphi_{i_n}(x_0)$$

for any $x_0 \in X$.

Definition 7.8.

(i) The set of infinite sequences $I^{\mathbb{N}}$ over the index set $I := \{1, \ldots, N\}$ is called *code space* of the iterated function system $\varphi_1, \ldots, \varphi_N$ (or of the attractor A).

(ii) The mapping $\pi : I^{\mathbb{N}} \to X$ defined as

$$\pi(i_1, i_2, \ldots) := \lim_{n \to \infty} \varphi_{i_1} \circ \cdots \circ \varphi_{i_n}(x_0)$$

(which does not depend on the choice of the starting point x_0) with $\pi(I^{\mathbb{N}}) = A$ is called *projection* of the code space onto the attractor.

Note that the projection $\pi : I^{\mathbb{N}} \to X$ in (ii) is determined for contractions $\varphi_1, \ldots, \varphi_N$ in arbitrary complete metric spaces (X, d) by the following arguments:

$$d(\varphi_{i_1} \circ \cdots \circ \varphi_{i_n}(x_0), \, \varphi_{i_1} \circ \cdots \circ \varphi_{i_n} \circ \varphi_{i_{n+1}} \circ \cdots \circ \varphi_{i_{n+m}}(x_0))$$
$$\leq r_{max}^n \, d(x_0, \varphi_{i_{n+1}} \circ \cdots \circ \varphi_{i_{n+m}}(x_0)).$$

Thus, $(\varphi_{i_1} \circ \cdots \circ \varphi_{i_n}(x_0))_{n=1}^{\infty}$ forms a Cauchy sequence, if the last distance remains bounded. In order to see this, we estimate

$$d(\varphi_{k_1} \circ \cdots \circ \varphi_{k_n}(x_0), \, x_0)$$
$$\leq \sum_{l=1}^{n} d(\varphi_{k_1} \circ \cdots \circ \varphi_{k_l}(x_0), \, \varphi_{k_1} \circ \cdots \circ \varphi_{k_{l-1}}(x_0))$$
$$\leq \sum_{l=1}^{\infty} r_{max}^{l-1} \, \max_{1 \leq i \leq N} d(\varphi_i(x_0), x_0) = const$$

for all code sequences $(k_1, k_2, \ldots)$.

We next provide the code space $I^{\mathbb{N}}$ with the *product topology* generated by the cylinder sets

$$\{(i_1, i_2, \ldots) \in I^{\mathbb{N}} : i_1 = k_1, \ldots, i_n = k_n\}, \quad k_1, \ldots, k_n \in I, \ n \in \mathbb{N}.$$

Then $I^{\mathbb{N}}$ as product of compact spaces I is again compact. By the last estimates from above the mapping $\pi : I^{\mathbb{N}} \to X$ is *bounded*. Moreover, it is *continuous* w.r.t. the product topology, since

$$d\left(\pi(k_1, k_2, \ldots), \, \pi(k_1, k_2, \ldots, k_n, \, i_{n+1}, i_{n+2}, \ldots)\right)$$
$$\leq r_{max}^n \, d\left(\pi(k_{n+1}, k_{n+2}, \ldots), \, \pi(i_{n+1}, i_{n+2}, \ldots)\right)$$
$$\leq r_{max}^n \cdot const, \quad \text{and} \quad r_{max} < 1.$$

Recall that the continuous image of a compact set is compact. Therefore the set $A \in \mathcal{K}$ in the general case may be introduced by

$$A := \pi(I^{\mathbb{N}}).$$

It is invariant under φ, since

$$\varphi(A) = \bigcup_{i_0=1}^{N} \varphi_{i_0}\left(\pi(I^{\mathbb{N}})\right)$$

$$= \bigcup_{i_0=1}^{N} \{\varphi_{i_0}(\pi(i_1, i_2, \ldots)) : (i_1, i_2, \ldots) \in I^{\mathbb{N}}\}$$

$$= \{\pi(i_0, i_1, \ldots) : (i_0, i_1, \ldots) \in I^{\mathbb{N}}\} = A.$$

Uniqueness of the compact fixed point A and its attractor property with respect to the Hausdorff distance may be shown as in the previous section. Consequently, we obtain the following *extension of Theorem 7.5.*

Theorem 7.9. *Let $\varphi_1, \ldots, \varphi_N$ be contractions of the complete metric space (X, d). Then there is a unique non-empty compact set A such that*

$$A = \varphi_1(A) \cup \cdots \cup \varphi_N(A).$$

Moreover,

$$\lim_{n \to \infty} d_H(\varphi^n(K_0), A) = 0$$

for any $K_0 \in \mathcal{K}$, where $\varphi(K) = \bigcup_{i=1}^{N} \varphi_i(K)$, $K \in \mathcal{K}$.

The representation $A = \pi(I^{\mathbb{N}})$ of the attractor provides the *second method for its generation*. The encoding sequences $(i_1, i_2, \ldots) \in I^{\mathbb{N}}$ are chosen "representative", i.e., *at random*, which means that the codes $i_1, i_2, \ldots$ are replaced by random versions: Let $\xi_1, \xi_2, \ldots$ be i.i.d. random variables with values in $\{1, \ldots, N\}$ and probabilities $\mathbb{P}\{\xi_n = i\} = r_i^D$, where r_i is the Lipschitz constant of φ_i. The number D is determined by the equation

$$\sum_{i=1}^{N} r_i^D = 1.$$

(The function $f(s) := \sum_{i=1}^{N} r_i^s$ is continuous and monotone decreasing, $f(0) = N \geq 2$ and $f(+\infty) = 0$. Thus, there is a unique $D > 0$ such that $\sum_{i=1}^{N} r_i^D = 1$.)

D is called the *fractal dimension* of the iterated function system (IFS) $\varphi_1, \ldots, \varphi_N$.

Later on we will show that the random variable

$$x_{\xi_1, \xi_2, \ldots} := \lim_{n \to \infty} \varphi_{\xi_1} \circ \cdots \circ \varphi_{\xi_n}(x_0)$$

is "uniformly" distributed on A provided the mappings $\varphi_1, \ldots, \varphi_N$ are similarities in a Euclidean space, satisfying an "almost non-overlapping" condition. Then, for sufficiently large n and large number of simulations the set of the simulation points $\varphi_{\xi_1} \circ \cdots \circ \varphi_{\xi_n}(x_0)$ forms a "good" approximation of A. The corresponding algorithm is faster than that of the first method mentioned in the previous section. See also Section 7.8 below for a still better method.

Note that the projection mapping π introduced in 7.8 is not necessarily biunique, since the images $\varphi_i(A)$ and $\varphi_j(A)$ for $i \neq j$ may overlap. However, the overlapping parts may be neglected as far as their D-dimensional Hausdorff measure is concerned:

Theorem 7.10. *Let $\varphi_1, \ldots, \varphi_N$ be contractions of a complete metric space with contraction ratios $r_1, \ldots, r_N$, and let D be given by the equation $\sum_{i=1}^{N} r_i^D = 1$. Then the corresponding attractor A has the following properties:*

(i) $\mathcal{H}^D(A) < \infty$, *i.e.,* $\dim_H A \leq D$.
(ii) $\mathcal{H}^D(\varphi_i(A) \cap \varphi_j(A)) = 0$, $i \neq j$.

Proof. (i) Using the representation

$$A = \bigcup_{i_1, \ldots, i_n \in I} \varphi_{i_1} \circ \cdots \circ \varphi_{i_n}(A)$$

and $|\varphi_{i_1} \circ \cdots \circ \varphi_{i_n}(A)| \leq r_{max}^n |A| < \delta$ for all sufficiently large n, we may choose the sets $\varphi_{i_1} \circ \cdots \circ \varphi_{i_n}(A)$ for a δ-covering of A. Then we obtain

$$\mathcal{H}_\delta^D(A) \leq \sum_{i_1, \ldots, i_n \in I} |\varphi_{i_1} \circ \cdots \circ \varphi_{i_n}(A)|^D$$

$$\leq \sum_{i_1, \ldots, i_n \in I} r_{i_1}^D \cdots r_{i_n}^D |A|^D = |A|^D ,$$

since

$$\sum_{i_1, \ldots, i_n \in I} r_{i_1}^D = \left(\sum_{i=1}^{N} r_i^D \right)^n = 1 .$$

Hence, for any $\delta > 0$, $\mathcal{H}_\delta^D(A) \leq |A|^D$, and therefore $\mathcal{H}^D(A) \leq |A|^D < \infty$, which yields (i). (ii) We have

$$\mathcal{H}^D(A) = \mathcal{H}^D \left(\bigcup_{i=1}^{N} \varphi_i(A) \right) \leq \sum_{i=1}^{N} \mathcal{H}^D(\varphi_i(A))$$

$$\leq \sum_{i=1}^{N} r_i^D \mathcal{H}^D(A) = \mathcal{H}^D(A)$$

by the behavior of $\mathcal{H}^D$ under Lipschitz mappings. This implies

$$\mathcal{H}^D\left(\bigcup_{i=1}^{N} \varphi_i(A)\right) = \sum_{i=1}^{N} \mathcal{H}^D(\varphi_i(A))$$

and, by additivity,

$$\mathcal{H}^D(\varphi_i(A) \cap \varphi_j(A))) = 0, \quad i \neq j.$$

$\square$

7.3 Dimensions of self-similar sets in Euclidean spaces

We now restrict to the special case, where the φ_i are contractive similarities $S_1, \ldots, S_N$ of $\mathbb{R}^d$ with the usual metric. Then the fractal dimension D given by $\sum_{i=1}^{N} r_i^D = 1$ is also called *similarity dimension*.

Proposition 7.11. *If A is the attractor of the system $S_1, \ldots, S_N$, $0 < \mathcal{H}^s(A) < \infty$, and $\mathcal{H}^s(S_i(A) \cap S_j(A)) = 0$, $i \neq j$, for some $s \leq D$, then $\dim_H A = s = D$.*

Proof. Clearly, $\dim_H A = s$. Furthermore,

$$\mathcal{H}^s(A) = \mathcal{H}^s\left(\bigcup_{i=1}^{N} S_i(A)\right) = \sum_{i=1}^{N} \mathcal{H}^s(S_i(A)) = \sum_{i=1}^{N} r_i^s \mathcal{H}^s(A).$$

Consequently, $\sum_{i=1}^{N} r_i^s = 1$, which yields $s = D$. $\square$

We already know that $\mathcal{H}^D(A) < \infty$ (cf. Theorem 7.10). A geometric condition guaranteeing $\mathcal{H}^D(A) > 0$ reads as follows:

Definition 7.12. The iterated function system $S_1, \ldots, S_N$ satisfies the *Open Set Condition* (OSC), if there exists a non-empty bounded open set $O \subset \mathbb{R}^d$ such that

$$\text{(i)} \quad \bigcup_{i=1}^{N} S_i(O) \subset O,$$

$$\text{(ii)} \quad S_i(O) \cap S_j(O) = \emptyset, \quad i \neq j.$$

Remark 7.13. All examples from 5.5 fulfill this condition: for O we can take the figures in the primary construction steps of the Cantor set, the Koch curve (second construction), the Sierpinski triangle and the Sierpinski carpet, respectively.

The following basic result was proved by [Moran (1946)] for $\mathbb{R}^1$ and in a more general context by [Hutchinson (1981)].

Theorem 7.14. *If $S_1, \ldots, S_N$ are contractive similarities of $\mathbb{R}^d$ with ratios $r_1, \ldots, r_N$ satisfying the Open Set Condition and D is given by $\sum_{i=1}^{N} r_i^D = 1$, then we have for the corresponding self-similar set A:*

$$\text{(i)} \quad 0 < \mathcal{H}^D(A) < \infty,$$
$$\text{(ii)} \quad \dim_H A = \dim_P A = \dim_E A = D.$$

Remark 7.15. The Open Set Condition is not restrictive in (i). By a result of [Schief (1994)], which uses [Bandt and Graf (1992)], for the fractal dimension D the condition $\mathcal{H}^D(A) > 0$ implies the existence of a bounded open set O fulfilling even the so-called *Strong Open Set Condition* (SOSC), i.e.,

$$\text{(OSC)} \quad \text{and} \quad O \cap A \neq \emptyset. \tag{7.1}$$

(This means that the attractor A is not concentrated on the boundary of the open set O.)

Proof of Theorem 7.14. Below we will construct a mass distribution μ in $\mathbb{R}^n$ with supp $\mu = A$ such that

$$a < \frac{\mu(B(x,r))}{r^D} < b, \quad x \in A, \quad 0 < r < |O|,$$

for some constants $a, b > 0$ (cf. 7.17). Then Theorem 3.7 (i) implies $\mathcal{H}^D(A) > 0$. (The known inequality $\mathcal{H}^D(A) < \infty$, see Theorem 7.10, also follows from 3.7 (ii).)

Furthermore, Proposition 5.10 implies $\overline{\dim}_E A \leq D$. Since $\dim_H A \leq \dim_P A \leq \overline{\dim}_E A$ we obtain the assertion. $\quad\square$

The *construction of the desired mass distribution μ on A* is similar as in the case of the Cantor set.

First note that $\overline{O}$ is compact and $\bigcup_{i=1}^{N} S_i(\overline{O}) \subset \overline{O}$. In view of Theorem 7.5 we get

$$A = \bigcap_{n=1}^{\infty} \bigcup_{i_1, \ldots, i_n \in \{1, \ldots, N\}} S_{i_1} \circ \cdots \circ S_{i_n}(\overline{O}).$$

In the case of the Cantor set the retracted sets $S_{i_1} \circ \cdots \circ S_{i_n}(\overline{O})$ in each step have got equal mass 2^{-n}. In the general case one could choose the masses N^{-n}. However, in this way the possibly different contraction ratios

$r_1, \dots, r_N$ are not taken into regard. It turns out that the right choice are the weights $r_{i_1}^D \cdots r_{i_n}^D$ for the disjoint sets $S_{i_1} \circ \cdots \circ S_{i_n}(\overline{O})$. This idea may be formalized in the language of the code space as follows.

Let ν_0 be the probability measure on $I = \{1, \dots, N\}$ with $\nu_0(\{i\}) = r_i^D$, $i \in I$, and

$$\nu := \overset{\infty}{\underset{i=1}{\otimes}} \nu_0$$

be the corresponding *product measure* on $I^{\mathbb{N}}$ provided with the product σ-algebra. Using the projection $\pi : I^{\mathbb{N}} \to \mathbb{R}^d$ with $\pi(i_1, i_2, \dots) = \lim_{n \to \infty} S_{i_1} \circ \cdots \circ S_{i_n}(x_0)$ and $\pi(I^{\mathbb{N}}) = A$ as above we choose the image measure of ν on $[\mathbb{R}^d, \mathfrak{B}(\mathbb{R}^d)]$:

Definition 7.16.

$$\mu := \nu \circ \pi^{-1}$$

By construction, $\mu(A) = \mu(\mathbb{R}^d) = 1$. (Below we will show that μ agrees with the normalized D-dimensional Hausdorff measure restricted to A.)

Theorem 7.17. *Under the conditions of Theorem 7.14 there exist some constants $a, b > 0$ such that the measure μ introduced in Definition 7.16 satisfies*

$$a \leq \frac{\mu(B(x, r))}{r^D} \leq b, \quad x \in A, \quad 0 < r < |O|.$$

In particular, $\operatorname{supp} \mu = A$.

Proof. *Lower estimate:* By construction,

$$\mu(S_{i_1} \circ \cdots \circ S_{i_n}(\overline{O})) \geq \nu(\{j \in I^{\mathbb{N}} : j_1 = i_1, \dots, j_n = i_n\}) = r_{i_1}^D \cdots r_{i_n}^D. \quad (7.2)$$

Further, for any $x \in A$ there exists $(i_1, i_2, \dots) \in I^{\mathbb{N}}$ with $x = \pi(i_1, i_2, \dots)$. For this sequence and $r < |O|$ let $\tau = \tau_r(i_1, i_2, \dots)$ be the smallest number satisfying

$$r_{i_1} \cdots r_{i_\tau} |O| < r. \quad (7.3)$$

Hence,

$$r_{i_1} \cdots r_{i_\tau} |O| \geq r_{i_1} \cdots r_{i_{\tau-1}} r_{min} |O| \geq r \, r_{min}, \quad (7.4)$$

where $r_{min} := \min_{1 \leq i \leq N} r_i$.

Condition (7.3) yields $S_{i_1} \circ \cdots \circ S_{i_\tau}(\overline{O}) \subset B(x, r)$. This together with (7.4) and (7.2) implies

$$\frac{\mu(B(x, r))}{r^D} \geq \frac{\mu(S_{i_1} \circ \cdots \circ S_{i_\tau}(\overline{O}))}{r_{i_1}^D \cdots r_{i_\tau}^D} \frac{r_{min}^D}{|O|^D} \geq \frac{r_{min}^D}{|O|^D} =: a,$$

i.e. the desired lower estimate.

Upper estimate: Cutting the sequences $(i_1, i_2, \dots) \in I^{\mathbb{N}}$ at the corresponding index τ introduced above, we turn to the following set of finite sequences in I of different lengths:

$$I_r := \{(i_1, \dots, i_\tau) : i_j \in I, \ r_{i_1} \cdots r_{i_\tau}|O| < r \leq r_{i_1} \cdots r_{i_{\tau-1}}|O|$$
$$(\leq r_{i_1} \cdots r_{i_\tau} r_{min}^{-1}|O|).$$

This set is sometimes called *Markov stop*.

Since

$$A = \pi(I^{\mathbb{N}}) = \bigcup_{I^{\mathbb{N}}} \bigcap_{n=1}^{\infty} S_{i_1} \circ \cdots \circ S_{i_n}(A)$$

$$= \bigcup_{I^{\mathbb{N}}} S_{i_1} \circ \cdots \circ S_{i_\tau} \left(\bigcap_{m=1}^{\infty} S_{i_{\tau+1}} \circ \cdots \circ S_{i_{\tau+m}}(A) \right)$$

$$= \bigcup_{I_r} S_{i_1} \circ \cdots \circ S_{i_\tau} \left(\bigcup_{I^{\mathbb{N}}} \bigcap_{m=1}^{\infty} S_{j_1} \circ \cdots \circ S_{j_m}(A) \right)$$

$$= \bigcup_{I_r} S_{i_1} \circ \cdots \circ S_{i_\tau}(A),$$

we obtain *self-similarity of A with respect to the Markov stop I_r*:

$$A = \bigcup_{I_r} S_{i_1} \circ \cdots \circ S_{i_\tau}(A). \tag{7.5}$$

Thus,

$$\frac{\mu(B(x,r))}{r^D} = \frac{\mu(B(x,r) \cap A)}{r^D}$$

$$= r^{-D} \mu \left(\bigcup_{I_r} B(x,r) \cap S_{i_1} \circ \cdots \circ S_{i_\tau}(A) \right)$$

$$\leq r^{-D} \nu \left((i_1, i_2, \dots) \in I^{\mathbb{N}} : (i_1, \dots, i_\tau) \in I_r' \right),$$

where $I_r' := \{(i_1, \dots, i_\tau) \in I_r : S_{i_1} \circ \cdots \circ S_{i_\tau}(\overline{O}) \cap B(x,r) \neq \emptyset\}$.

By definition of ν the last expression is equal to

$$r^{-D} \sum_{(i_1, \dots, i_\tau) \in I_r'} r_{i_1}^D \cdots r_{i_\tau}^D \leq r^{-D} r^D |O|^{-D} \operatorname{card}(I_r') = |O|^{-D} \operatorname{card}(I_r'),$$

where card means the cardinality of a finite set. Therefore it is enough to show the existence of some $c > 0$ such that $\operatorname{card}(I_r') < c$ for $0 < r < |O|$.

This number of elements may be estimated by comparing volumes in $\mathbb{R}^d$: The sets

$$S_{i_1} \circ \cdots \circ S_{i_\tau}(O), \ (i_1, \dots, i_\tau) \in I_r',$$

are open and disjoint by (OSC) and the definition of τ. Since $r_{i_1} \cdots r_{i_\tau} > r_{min}|O|^{-1}r$, each of them contains an open ball $B_{i_1,\ldots,i_\tau}$ of radius $(r_{min}|O|^{-1}r)r_o$, where r_o denotes the radius of the inball of O. Furthermore, the diameters of the above sets are less than r and therefore $S_{i_1} \circ \cdots \circ S_{i_\tau}(O) \subset B(x,2r)$, provided $S_{i_1} \circ \cdots \circ S_{i_\tau}(\overline{O}) \cap B(x,r) \neq \emptyset$. Consequently,

$$\omega_d(r_{min}|O|^{-1}r_o)^d r^d \operatorname{card}(I'_r) \leq \sum_{(i_1,\ldots,i_\tau)\in I'_r} \mathcal{L}^d(S_{i_1} \circ \cdots \circ S_{i_\tau}(O))$$

$$\leq \mathcal{L}^d(B(x,2r)) = \omega_d(2r)^d\,,$$

where ω_d is the volume of the unit ball in $\mathbb{R}^d$. Then we get the desired estimate

$$\operatorname{card}(I'_r) \leq \frac{2^d|O|^d}{r_{min}^d\, r_o^d} =: c, \quad 0 < r < |O|\,.$$

$\square$

7.4   Extension to metric spaces

In the previous section the Euclidean structure of the metric space $\mathbb{R}^d$ has been used only at one place. For the upper estimate

$$\frac{\mu(B(x,r))}{r^D} < b, \quad x \in A, \quad 0 < r < |O|\,,$$

in Theorem 7.17 we have applied the method of comparing volumes of balls in $\mathbb{R}^d$. What we actually have exploited is the following property of the space $\mathbb{R}^d$, which characterizes a class of more general metric spaces.

Definition 7.18. The metric space (Y, d) is called β-*space*, if for any $0 < \beta < 1$ there are constants $C(\beta), D(\beta) > 0$ such that for the maximal number $N_r(\beta)$ of disjoint closed balls with radius βr and centers in a ball of radius r the following condition is fulfilled:

$$N_r(\beta) < C(\beta), \quad 0 < r < D(\beta)\,.$$

Note that the metric spaces of *homogeneous type* used in harmonic analysis are special examples: They are determined by $C(\beta) := \operatorname{const} \beta^{-\gamma}$, for some $\gamma > 0$ and $D(\beta) := +\infty$.

Exercise 7.2.

(i) In the definition of a β-space the number $N_r(\beta)$ may be replaced
by the maximal number of points in Y whose mutual distances are
greater or equal to $\beta\,r$.
(Hint: for $d(x,y) \geq \beta\,r$ and $x,y \in B(z,r)$ the balls $B(x,\beta\,r/3)$ and
$B(y,\beta\,r/3)$ are contained in $B(z,(1+\beta/3)r)$.)

(ii) Any subspace (Y',d) of a β-space (Y,d) is a β-space, too.

Let now (X,d) be a complete metric space and A be the attractor of the
contractive similarities $S_1,\ldots,S_N$. Recall that the latter fulfills the Strong
Open Set Condition (SOSC), if they satisfy (OSC) for some O such that
$O \cap A \neq \emptyset$.

Analogous arguments as in Section 7.3 (where in the proof of Theorem
7.17 the set O is now to be replaced by $O \cap A$) lead to the following.

Theorem 7.19. *Let $S_1,\ldots,S_N$ be contractive similarities of the complete
metric space (X,d) satisfying the Strong Open Set Condition for the attractor A, and let (A,d) be a β-space. Then the assertions of Theorems 7.14
and 7.17 remain valid.*

Remarks 7.1.

(i) Again, the conditions of Theorem 7.19 are not restricting for
$\mathcal{H}^D(A) > 0$. [Schief (1996)] has shown that $\mathcal{H}^D(A) > 0$ implies
that A is a β-space and (SOSC) is fulfilled.

(ii) The very same arguments as in Section 7.3 show that the assertions
of Theorems 7.14 and 7.17 remain valid, if (X,d) is a *complete
β-space* and $S_1,\ldots,S_N$ satisfy (OSC).

(iii) A sufficient condition for these statements in *arbitrary complete
metric spaces* is the *Strong Separation Condition* (SSC):

$$S_i(A) \cap S_j(A) = \emptyset, \quad i \neq j.$$

(Here, the proof of 7.17 is easier when taking instead of $B(x,r)$ the
ball $B(x,\rho x)$ for a sufficiently small constant $\rho > 0$ determined by
the minimal distance between the sets $S_i(A)$ and $S_j(A)$, $i \neq j$.)

(iv) Then (ii) and (iii) may be considered as special cases of Theorem 7.19: The corresponding conditions lead to $\mathcal{H}(A) > 0$ and by
Schief's result to those in 7.19.

7.5   Invariant measures

The measure μ constructed in Sections 7.3 and 7.4 as image measure of the product measure ν on the code space under the projection π agrees with the normalized D-dimensional Hausdorff measure on the self-similar set A:

Theorem 7.20. *Let $S_1, \ldots, S_N$ be contractive similarities of the complete metric space (X, d) with contraction ratios $r_1, \ldots, r_N$ and $\sum_{i=1}^{N} r_i^D = 1$. Suppose that $S_1, \ldots, S_N$ satisfy (SOSC) (7.1) and that the self-similar set A is a β-space. Let $\pi : \{1, \ldots, N\}^{\mathbb{N}} \to X$ be the projection mapping onto A and $\nu = \bigotimes_{i=1}^{\infty} \nu_0$ be the product measure on the code space with $\nu_0(i) = r_i^D$, $i = 1, \ldots, N$. Then we have the following:*

(i) *There is a unique normalized measure μ on $[X, \mathfrak{B}(X)]$ such that*

$$\mu = \sum_{i=1}^{N} r_i^D \, \mu \circ S_i^{-1} \, .$$

(ii) $\mu = \nu \circ \pi^{-1} = \mathcal{H}^D(A)^{-1} \, \mathcal{H}^D \llcorner A$, *and* $\operatorname{supp} \mu = A$.

Definition 7.21. Property (i) is called *self-similarity of μ*.

Remark 7.22. In view of Remarks 7.1 the conditions of Theorem 7.20 are fulfilled, if (X, d) is a complete β-space, in particular if $X = \mathbb{R}^d$, and if the similarities satisfy (OSC).

Proof of Theorem 7.20. By Theorem 7.19 we have $0 < \mathcal{H}^D(A) < \infty$. Set $\mu := \mathcal{H}^D(A)^{-1} \mathcal{H}^D \llcorner A$. Then we get, in view of Theorem 7.10 (ii) and the scaling property of the Hausdorff measure, for any Borel set B,

$$\mathcal{H}^D(A) \, \mu(B) = \mathcal{H}^D(A \cap B) = \sum_{i=1}^{N} \mathcal{H}^D(S_i(A) \cap B)$$

$$= \sum_{i=1}^{N} \mathcal{H}^D(S_i(A \cap S_i^{-1}(B))) = \sum_{i=1}^{N} r_i^D \, \mathcal{H}^D(A \cap S_i^{-1}(B))$$

$$= \mathcal{H}^D(A) \sum_{i=1}^{N} r_i^D \, \mu(S_i^{-1}(B)) \, ,$$

which yields

$$\mu = \sum_{i=1}^{N} r_i^D \, \mu \circ S_i^{-1} \, .$$

Below we will show that the self-similarity of a normalized measure μ implies

$$\mu = \nu \circ \pi^{-1}. \tag{7.6}$$

This and the above arguments yield the equalities in (ii), in particular, uniqueness. Moreover, in view of the version of Theorem 7.17 in our metric space (cf. the arguments before Theorem 7.20) we have $\text{supp}(\nu \circ \pi^{-1}) = A$.

In order to show (7.6) we first consider closed subsets F of X and

$$F_n := \{(i_1, \ldots, i_n) \in I^n : S_{i_1} \circ \cdots \circ S_{i_n}(A) \cap F \neq \emptyset\}.$$

Then we obtain

$$\widetilde{F}_n := \{(i_1, i_2, \ldots) \in I^{\mathbb{N}} : (i_1, \ldots, i_n) \in F_n\} \downarrow \pi^{-1}(F)$$

as $n \to \infty$ and

$$\mu(F) = \sum_{(i_1, \ldots, i_n) \in I^n} r_{i_1}^D \cdots r_{i_n}^D \, \mu\left((S_{i_1} \circ \cdots \circ S_{i_n})^{-1}(F)\right)$$

$$\leq \sum_{(i_1, \ldots, i_n) \in F_n} r_{i_1}^D \cdots r_{i_n}^D = \nu(\widetilde{F}_n),$$

which tends to $\nu \circ \pi^{-1}(F)$ as $n \to \infty$, i.e.,

$$\mu(F) \leq \nu \circ \pi^{-1}(F).$$

Finally, let B be an arbitrary Borel set. By Theorem 1.24 for any $\varepsilon > 0$ there exists a closed $F \subset B$ such that $\mu(B \setminus F) < \varepsilon$. Consequently,

$$\mu(B) \leq \mu(F) + \mu(B \setminus F) \leq \nu \circ \pi^{-1}(F) + \varepsilon \leq \nu \circ \pi^{-1}(B) + \varepsilon,$$

and therefore

$$\mu(B) \leq \nu \circ \pi^{-1}(B) \quad \text{for any Borel set } B.$$

In particular,

$$\mu(X \setminus B) \leq \nu(X \setminus \pi^{-1}(B)), \text{i.e.,}$$

$$1 - \mu(B) \leq 1 - \nu \circ \pi^{-1}(B),$$

and thus

$$\mu(B) \geq \nu \circ \pi^{-1}(B).$$

Together with the inverse inequality obtained above we infer

$$\mu(B) = \nu \circ \pi^{-1}(B), \quad B \in \mathfrak{B}(X),$$

which completes the proof. $\qquad\qquad\qquad\qquad\qquad\qquad\qquad\qquad\Box$

Exercise 7.3. The assertion of Theorem 7.20 remains valid for Borel regular outer measures μ on X and the extension of ν to an outer measure ν^* (cf. 1.3).

If we replace the above similarities by contractions and the marginal measure ν_0 in the product measure $\nu = \bigotimes_{i=1}^{\infty} \nu_0$ on the code space $I^{\mathbb{N}}$ by a probability distribution with arbitrary weights $p_i > 0$, i.e.,

$$\nu_0(\{i\}) = p_i,\ p_i > 0,\ i = 1,\dots,N,\quad \sum_{i=1}^{N} p_i = 1,$$

then we turn to the following *extension of Theorem* 7.20, except of the relationship to Hausdorff measure. (Recall that by A we denoted the unique compact set invariant with respect to the IFS under consideration.)

Theorem 7.23. *Let* $\varphi_1,\dots,\varphi_N$ *be contractions of the complete metric space* (X, d) *and* $\mu := \nu \circ \pi^{-1}$ *with* ν *chosen as above. Then* μ *is the unique normalized measure on* $[X, \mathfrak{B}(X)]$ *which is invariant with respect to the iterated function system* $\varphi_1,\dots,\varphi_N$ *with weights* $p_1,\dots,p_N$, *i.e.,*

$$\mu = \sum_{i=1}^{N} p_i\, \mu \circ \varphi_i^{-1}.$$

Moreover, $\operatorname{supp}\mu = \pi(I^{\mathbb{N}}) = A$.

Proof. For $\mu = \nu \circ \pi^{-1}$ we get in view of the product structure of ν

$$\sum_{i=1}^{N} p_i\, \mu \circ \varphi_i^{-1} = \sum_{i=1}^{N} p_i\, \nu(\{j : \varphi_i(\pi(j_1, j_2,\dots)) \in (\cdot)\})$$

$$= \sum_{i=1}^{N} p_i\, \nu(\{\pi(i, j_1, j_2,\dots) \in (\cdot)\}) = (\nu_0 \times \nu) \circ \pi^{-1} = \nu \circ \pi^{-1} = \mu,$$

i.e., μ satisfies the desired equation. Any other normalized measure with this invariance property agrees with μ by similar arguments as in the proof of Theorem 7.20.

In order to prove $\operatorname{supp}\mu = A$ first note that by construction $\operatorname{supp}\mu \subset A$. Therefore it suffices to show that for all $x \in A$ and $r > 0$ we have $\mu(B(x,r)) > 0$. The proof is similar as in the special where $p_i = r_i^D$: Recall that for any $x \in A$ there is a code sequence $(i_1, i_2,\dots) \in I^{\mathbb{N}}$ such that $\pi(i_1, i_2,\dots) = x$. Furthermore, for any $r > 0$ there exists some n with $r_{i_1} \cdots r_{i_n}|A| < r$, where $r_i < 1$ denotes the Lipschitz constants of φ_i, $i = 1,\dots,N$. Then we get $\varphi_{i_1} \circ \cdots \circ \varphi_{i_n}(A) \subset B(x,r)$ and therefore

$$\mu(B(x,r)) \geq \mu(\varphi_{i_1} \circ \cdots \circ \varphi_{i_n}(A))$$

$$\geq \nu(\{(j_1, j_2,\dots) \in I^{\mathbb{N}} : j_1 = i_1,\dots,j_n = i_n\}) = p_{i_1} \cdots p_{i_n} > 0,$$

since $\mu = \nu \circ \pi^{-1}$. $\qquad\square$

In the special case of similarities μ is also called *self-similar measure with weights* p_i.

In the rest of this section we will show that in the case of Polish spaces (X, d) the measure μ may be interpreted as *attractor* of the iterated function system $\varphi_1, \ldots, \varphi_N$ on the space $\mathcal{P}_b(X)$ of *normalized measures on* $[X, \mathfrak{B}(X)]$ *with bounded supports* provided with a suitable metric. Let

$$\mathrm{Lip}_1(X) := \{f : X \to \mathbb{R} : |f(x) - f(y)| \le d(x, y), \ x, y \in X\}.$$

be the space of *non-expanding* Lipschitz functions on (X, d).

Definition 7.24. If (X, d) is a complete separable metric space, then the *Monge-Kantorovich metric* on $\mathcal{P}_b(X)$ is defined as

$$\Delta_{MK}(P, Q) := \sup \left\{ \int f \, dP - \int f \, dQ : f \in \mathrm{Lip}_1(X) \right\}, \quad P, Q \in \mathcal{P}_b(X).$$

Proposition 7.25. $\Delta_{MK}(P, Q) < \infty$, *for any* $P, Q \in \mathcal{P}_b(X)$.

Proof. Suppose that $\mathrm{supp}\, P$ and $\mathrm{supp}\, Q$ are contained in the ball $B(y, r)$ and take $f \in L_1(X)$. Since $\int f(y) P(dx) = \int f(y) Q(dx) = f(y)$, we get

$$\int f \, dP - \int f \, dQ$$

$$= \int (f(x) - f(y)) P(dx) - \int (f(x) - f(y)) Q(dx)$$

$$\le \int |f(x) - f(y)| P(dx) + \int |f(x) - f(y)| Q(dx)$$

$$\le \int_{\mathrm{supp}\, P} d(x, y) P(dx) + \int_{\mathrm{supp}\, Q} d(x, y) Q(dx) \le 4r,$$

using the non-expanding property of the functions f and that $d(x, y) \le 2r$ under the integrals. $\qquad\square$

Exercise 7.4. Δ_{MK} is a metric on $\mathcal{P}_b(X)$.

We now introduce for any iterated function system of contractions $\varphi_1, \ldots, \varphi_N$ on X with weights $p_1, \ldots, p_N$ a transformation Φ on $\mathcal{P}_b(X)$, which plays the role of the mapping φ on the space $\mathcal{K} = \mathcal{K}(X)$ of compact subsets of X used in Sections 7.1–7.3.

Definition 7.26. $\Phi : \mathcal{P}_b(X) \to \mathcal{P}_b(X)$ is given by

$$\Phi(P) := \sum_{i=1}^{N} p_i \, P \circ \varphi_i^{-1}.$$

Proposition 7.27. *Let $\varphi_1,\ldots,\varphi_N$ be contractions on the complete separable metric space (X,d) with Lipschitz constants $r_1,\ldots,r_N$. Then Φ is a contraction on the metric space $(\mathcal{P}(X),\Delta_{MK})$ whose Lipschitz constant does not exceed r_{max}.*

Proof.

$$\Delta_{MK}(\Phi(P),\Phi(Q)) = \sup\left\{ \int f\,d(\Phi(P)) - \int f\,d(\Phi(Q)) : f \in \mathrm{Lip}_1(X)\right\}$$

$$= \sup\left\{ \sum_{i=1}^{N} p_i \int f\circ\varphi_i\,dP - \sum_{i=1}^{N} p_i \int f\circ\varphi_i\,dQ : f \in \mathrm{Lip}_1(X)\right\}.$$

Denote $\tilde{f}(x) := \sum_{i=1}^{N} p_i\, f(\varphi_i(x))$. Then $\tilde{f}$ satisfies

$$|\tilde{f}(x) - \tilde{f}(y)|$$
$$\leq \sum_{i=1}^{N} p_i|f(\varphi_i(x)) - f(\varphi_i(y))| \leq \sum_{i=1}^{N} p_i d(\varphi_i(x),\varphi_i(y))$$
$$\leq \sum_{i=1}^{N} p_i\, r_i\, d(x,y) \leq r_{max}\, d(x,y)\,,$$

i.e., $r_{max}^{-1}\tilde{f} \in \mathrm{Lip}_1(X)$. Therefore we obtain $r_{max}\,\Delta_{MK}(P,Q)$ as an upper estimate for the above expression. □

The *attractor property* of the Φ-invariant measure μ from Theorem 7.23 is an immediate consequence:

Corollary 7.28. *If (X,d) is a complete separable metric space, then for any starting measure $\mu_0 \in \mathcal{P}_b(X)$ we have the convergence*

$$\lim_{n\to\infty} \Delta_{MK}(\Phi^n(\mu_0),\mu) = 0\,.$$

Proof.

$$\Delta_{MK}(\Phi^n(\mu_0),\mu) = \Delta_{MK}(\Phi^n(\mu_0),\Phi^n(\mu)) \leq r_{max}^n\,\Delta_{MK}(\mu_0,\mu)\,,$$

which tends to 0 as $n\to\infty$. □

7.6 Dimensions of invariant measures

Next we will derive an upper estimate for the pointwise dimensions (recall Definition 6.1) of the invariant measure $\mu = \nu\circ\pi^{-1}$ from Theorem 7.23 in the general case.

Theorem 7.29. *Let $\varphi_1,\ldots,\varphi_N$ be contractions of the complete metric space (X,d) with Lipschitz constants $r_1,\ldots,r_N$. Then the invariant measure μ as above with weights $p_1,\ldots,p_N$ satisfies*

$$\overline{\dim}\mu(x) \leq \frac{\sum\limits_{i=1}^{N} p_i \log p_i}{\sum\limits_{i=1}^{N} p_i \log r_i}$$

at μ-almost all $x \in X$.

Proof. If $x \in A$ then there exists an encoding sequence $(i_1,i_2,\ldots)$ such that $x = \pi(i_1,i_2,\ldots)$ for the projection π. Similarly as before for $0 < r < |A|$ define the index $\tau = \tau((i_1,i_2,\ldots),r)$ by

$$r_{i_1}, r_{i_2}\cdots r_{i_\tau}\,|A| < r \leq r_{i_1}, r_{i_2}\cdots r_{i_{\tau-1}}\,|A| \quad (\leq r_{min}^{-1}\, r_{i_1}, r_{i_2}\cdots r_{i_\tau}\,|A|).$$

Then we obtain $B(x,r) \supset \varphi_{i_1} \circ \cdots \circ \varphi_{i_\tau}(A)$, hence,

$$\mu(B(x,r)) \geq \mu(\varphi_{i_1} \circ \cdots \circ \varphi_{i_\tau}(A))$$
$$\geq \nu(\{(j_1,j_2,\ldots) : j_1 = i_1,\ldots,j_\tau = i_\tau\}) = p_{i_1} \cdots p_{i_\tau}.$$

This implies for $r < \min(1,|A|)$,

$$\frac{\log \mu(B(x,r))}{\log r} \leq \frac{\log(p_{i_1}\cdots p_{i_\tau})}{\log \tau + \log|A| - \log r_{min}}$$

and therefore

$$\mathrm{limsup}_{r\to 0}\, \frac{\log \mu(B(x,r))}{\log r} \leq \mathrm{limsup}_{n\to\infty}\, \frac{\log(p_{i_1}\cdots p_{i_n})}{\log(r_{i_1}\cdots r_{i_n})}$$

$$= \frac{\lim_{n\to\infty} \frac{1}{n}\sum_{k=1}^{n}\log p_{i_k}}{\lim_{n\to\infty} \frac{1}{n}\sum_{k=1}^{n}\log r_{i_k}} = \frac{\sum_{i=1}^{N} p_i \log p_i}{\sum_{i^1}^{N} p_i \log r_i} =: c$$

for μ-almost all x, since the last two limits exist and are equal to the corresponding constants for ν-almost all $(i_1,i_2,\ldots) \in I^{\mathbb{N}}$. (Here we have used the strong law of large numbers for a random sequence $(\xi_1,\xi_2,\ldots)$ with distribution ν, i.e.,

$$\lim_{n\to\infty} \frac{1}{n}\sum_{k=1}^{n} h(\xi_k) = \mathbb{E}h(\xi_1)$$

with probability 1, where $h(i) = \log p_i$ and $h(i) = \log r_i$, respectively.) More precisely, if $M := \{x \in A : \overline{\dim}\mu(x) \leq c\}$ and M' is the set of sequences in $I^{\mathbb{N}}$ with the above convergence properties, then the above estimates show that $M' \subset \pi^{-1}(M)$. Using that $\nu(M') = 1$, we obtain $\mu(M) = \nu \circ \pi^{-1}(M) = 1$, since ν has mass 1. $\qquad\square$

In order to find a *lower estimate*, below we will restrict to the *case of similarities* $S_1, \ldots, S_N$. As an auxiliary tool we need the following result, which may be formulated in a more general context.

Note that two sequences $(i_1, \ldots, i_r)$ and $(j_1, \ldots, j_s)$ over I are said to be *incomparable*, if non of both agrees with the first part of the other one.

Proposition 7.30. *Let $\varphi_1, \ldots, \varphi_N$ be contractive injections of the complete metric space (X, d) satisfying the Strong Open Set Condition with respect to the set O. Then we have the following for the corresponding invariant measure μ with weights $p_1, \ldots, p_N$.*

(i) $\mu(\partial O) = 0$.

(ii) $\mu(\varphi_{i_1} \circ \cdots \circ \varphi_{i_r}(A) \cap \varphi_{j_1} \circ \cdots \circ \varphi_{j_s}(A)) = 0$
   for all incomparable sequences $(i_1, \ldots, i_r)$ and $(j_1, \ldots, j_s)$.

(iii) $\mu(\varphi_{i_1} \circ \cdots \circ \varphi_{i_n}(A)) = \mu(\varphi_{i_1} \circ \cdots \circ \varphi_{i_n}(\overline{O})) = p_{i_1} \cdots p_{i_n}$
   for all finite sequences $(i_1, \ldots, i_n)$.

Proof. (i) Since $A \cap O \neq \emptyset$ there is a finite sequence $(k_1, \ldots, k_m) \in I^m$ such that

$$\varphi_{k_1} \circ \cdots \varphi_{k_m}(A) \subset \varphi_{k_1} \circ \cdots \varphi_{k_m}(\overline{O}) \subset O.$$

Therefore $\mu(O) \geq \mu(\varphi_{k_1} \circ \cdots \varphi_{k_m}(A)) \geq p_{k_1} \cdots p_{k_m} > 0$. Furthermore, by (OSC) and invariance of μ we get

$$\mu(O) \geq \sum_{(i_1,\ldots,i_m)\in I^m} \mu(\varphi_{i_1} \circ \cdots \circ \varphi_{i_m}(O))$$

$$= \sum_{\substack{(i_1,\ldots,i_m)\in I^m \\ (j_1,\ldots,j_m)\in I^m}} p_{j_1} \cdots p_{j_m} \, \mu\big((\varphi_{j_1} \circ \cdots \circ \varphi_{j_m})^{-1}(\varphi_{i_1} \circ \cdots \circ \varphi_{i_m})(O)\big)$$

$$\geq \sum_{(i_1,\ldots,i_m)\in I^m} p_{i_1} \cdots p_{i_m} \, \mu(O) = \mu(O).$$

(In the last estimate we have omitted the sequences with $(i_1, \ldots, i_m) \neq (j_1, \ldots, j_m)$. Moreover, $\varphi_j^{-1} \circ \varphi_j = $ identity, since the mappings φ_j are supposed to be injective.) Hence,

$$\mu(O) = \sum_{(i_1,\ldots,i_m)\in I^m} \mu(\varphi_{i_1} \circ \cdots \circ \varphi_{i_m}(O)). \tag{7.7}$$

This means that

$$\mu\bigg(O \setminus \bigcup_{(i_1,\ldots,i_m)\in I^m} \varphi_{i_1} \circ \cdots \circ \varphi_{i_m}(O)\bigg) = 0.$$

Using
$$\varphi_{i_1} \circ \cdots \circ \varphi_{i_m}(O) \cap \varphi_{k_1} \circ \cdots \circ \varphi_{k_m}(\overline{O}) = \emptyset,$$
if $(i_1, \ldots, i_m) \neq (k_1, \ldots, k_m)$, and that $\varphi_{k_1} \circ \cdots \circ \varphi_{k_m}(\overline{O}) \subset O$ we infer from this
$$\mu(\varphi_{k_1} \circ \cdots \circ \varphi_{k_m}(\overline{O}) \setminus \varphi_{k_1} \circ \cdots \circ \varphi_{k_m}(O))$$
$$= \mu\left(\varphi_{k_1} \circ \cdots \circ \varphi_{k_m}(\overline{O}) \setminus \bigcup_{(i_1,\ldots,i_m)\in I^m} \varphi_{i_1} \circ \cdots \circ \varphi_{i_m}(O) \right) = 0.$$

Further,
$$\mu(\varphi_{k_1} \circ \cdots \circ \varphi_{k_m}(\overline{O}) \setminus \varphi_{k_1} \circ \cdots \circ \varphi_{k_m}(O)) = \mu(\varphi_{k_1} \circ \cdots \circ \varphi_{k_m}(\partial O))$$
$$= \sum_{(i_1,\ldots,i_m)\in I^m} p_{i_1} \cdots p_{i_m} \, \mu((\varphi_{1_1} \circ \cdots \circ \varphi_{i_m})^{-1}(\varphi_{k_1} \circ \cdots \circ \varphi_{k_m})(\partial O))$$
$$\geq p_{k_1} \cdots p_{k_m} \, \mu(\partial O),$$
which yields $\mu(\partial O) = 0$, i.e., assertion (i).

(ii)　Since $\operatorname{supp} \mu \subset A \subset \overline{O}$ we infer $\mu(X \setminus O) = 0$.

Let now $(i_1, \ldots, i_r)$ and $(j_1, \ldots, j_s)$ be incomparable. By (OSC) we obtain
$$\varphi_{i_1} \circ \cdots \circ \varphi_{i_r}(\overline{O}) \cap \varphi_{j_1} \circ \cdots \circ \varphi_{j_s}(O) = \emptyset,$$
i.e.,
$$(\varphi_{j_1} \circ \cdots \circ \varphi_{j_s})^{-1}(\varphi_{i_1} \circ \cdots \circ \varphi_{i_r})(\overline{O}) \cap O = \emptyset$$
and thus, using $\mu(X \setminus O) = 0$,
$$\mu((\varphi_{j_1} \circ \cdots \circ \varphi_{j_s})^{-1}(\varphi_{i_1} \circ \cdots \circ \varphi_{i_r})(\overline{O})) = 0.$$
From this we conclude for $r = s$
$$\mu(\varphi_{i_1} \circ \cdots \circ \varphi_{i_r}(\overline{O}))$$
$$= \sum_{(j_1,\ldots,j_r)\in I^r} p_{j_1} \cdots p_{j_r} \, \mu((\varphi_{j_1} \circ \cdots \circ \varphi_{j_r})^{-1}(\varphi_{i_1} \circ \cdots \circ \varphi_{i_r})(\overline{O})$$

$$\tag{7.8}$$

$$= p_{i_1} \cdots p_{i_r} \, \mu(\overline{O}) = p_{i_1} \cdots p_{i_r}.$$
For incomparable $(i_1, \ldots, i_r)$, $(j_1, \ldots, j_s)$ there exists $n \leq \min(r, s)$ such that $i_1 = j_1, \ldots, i_{n-1} = j_{n-1}$ and $i_n \neq j_n$. (7.7) and (7.8) imply
$$\mu(\overline{O}) = \mu(O) = \sum_{(l_1,\ldots,l_n)\in I^n} \mu(\varphi_{l_1} \circ \cdots \circ \varphi_{l_n}(O))$$
$$\leq \sum_{(l_1,\ldots,l_n)\in I^n} \mu(\varphi_{l_1} \circ \cdots \circ \varphi_{l_n}(\overline{O}))$$
$$= \sum_{(l_1,\ldots,l_n)\in I^n} p_{l_1} \cdots p_{l_n} \, \mu(\overline{O}) = \mu(\overline{O}).$$

Hence,

$$\mu(\overline{O}) = \sum_{(l_1,\dots,l_n)\in I^n} \mu(\varphi_{l_1} \circ \cdots \circ \varphi_{l_n}(\overline{O})),$$

which yields, in particular,

$$\mu\big((\varphi_{j_1} \circ \cdots \circ \varphi_{j_n})(\overline{O}) \cap (\varphi_{i_1} \circ \cdots \circ \varphi_{i_n})(\overline{O})\big) = 0,$$

for our incomparable sequences $(i_1,\dots,i_r)$ and $(j_1,\dots,j_s)$.

Finally use that

$$\mu(\varphi_{i_1} \circ \cdots \circ \varphi_{i_r}(A) \cap \varphi_{j_1} \circ \cdots \circ \varphi_{j_s}(A))$$
$$\leq \mu(\varphi_{i_1} \circ \cdots \circ \varphi_{i_n}(\overline{O}) \cap \varphi_{j_1} \circ \cdots \circ \varphi_{j_n}(\overline{O})) = 0.$$

Thus, (ii) is proved.

(iii) This follows from the equations (7.8) together with the same arguments, if we replace there $\overline{O}$ by the set A. $\qquad\square$

We now turn back to the problem of lower estimates for the pointwise dimensions of the measure μ. Let $\varphi_1 := S_1, \dots, \varphi_N$ be similarities of the complete metric space (X, d) with contraction ratios $r_1, \dots, r_N$ satisfying (SOSC) for some O, and $\mu = \sum_{i=1}^N p_i\, \mu \circ S_i^{-1}$ is the invariant measure with weights $p_i > 0$ and $\mu(X) = 1$.

By the last proposition μ is concentrated in O and the set of points $x \in A \cap O$ with unique coding $x = \pi(i_1, i_2, \dots)$ is of full measure. For such x we denote

$$\delta(x) = \delta(\pi(i_1, i_2, \dots)) := d(x, \partial O) \quad \text{and}$$
$$\delta_n(x) = \delta_n(\pi(i_1, i_2, \dots)) := d\big(x, \partial(S_{i_1} \circ \cdots \circ S_{i_n}(O))\big), \quad n \in \mathbb{N},$$

i.e., $\delta_n(x)$ is the *distance from the point x to the boundary of the contracted copy of O in the n-th iteration step containing x*. Since we are dealing with similarities we obtain

$$\delta_n(\pi(i_1, i_2, \dots)) = r_{i_1} \cdots r_{i_n}\, \delta(\pi(i_{n+1}, i_{n+2}, \dots)).$$

Given $r > 0$ we now choose n so that

$$\delta_n(x) \leq r < \delta_{n-1}(x)$$

and infer $B(x, r) \subset S_{i_1} \circ \cdots \circ S_{i_{n-1}}(\overline{O})$. Thus, by Proposition 7.30 (iii),

$$\mu(B(x, r)) \leq \mu(S_{i_1} \circ \cdots \circ S_{i_{n-1}}(\overline{O})) = p_{i_1} \cdots p_{i_{n-1}} \leq p_{i_1} \cdots p_{i_n}(p_{min})^{-1}.$$

Then we obtain for $0 < r < 1$ and such n,

$$\frac{\log \mu(B(x,r))}{\log r} \geq \frac{\log(p_{i_1} \cdots p_{i_n}) - \log p_{max}}{\log(r_{i_1} \cdots r_{i_n}) + \log \delta(\pi(i_{n+1}, i_{n+2}, \dots))}$$

$$= \frac{\frac{1}{n} \sum_{k=1}^{n} \log p_{i_k} - \frac{1}{n} \log p_{min}}{\frac{1}{n} \sum_{k=1}^{n} \log r_{i_k} + \frac{1}{n} \log \delta(\pi(i_{n+1}, i_{n+2}, \dots))} .$$

Below we will show that

$$\lim_{n \to \infty} \frac{1}{n} \log \delta(\pi(i_{n+1}, i_{n+2}, \dots)) = 0 \tag{7.9}$$

for ν-almost all $(i_1, i_2, \dots) \in I^{\mathbb{N}}$. Therefore the strong law of large numbers leads to

$$\liminf_{r \to 0} \frac{\log \mu(B(x,r))}{\log r} \geq \frac{\sum_{i=1}^{N} p_i \log p_i}{\sum_{i=1}^{N} p_i \log r_i}$$

for μ-almost all x which yields together with 7.6.1 the main result of this section (cf. Theorem 7.6.5 below).

In order to prove (7.9) we will apply Birkhoff's ergodic theorem (see, e.g., [Walters (1982)]) to the space $(I^{\mathbb{N}}, \nu)$ and the ergodic left shift operator $\Theta : I^{\mathbb{N}} \to I^{\mathbb{N}}$ with $\Theta(i_1, i_2, \dots) := (i_2, i_3, \dots)$. It states that for any ν-integrable function $f : I^{\mathbb{N}} \to \mathbb{R}$ we have

$$\lim_{n \to \infty} \frac{1}{n} \sum_{k=1}^{n} f(\Theta^k(i_1, i_2, \dots)) = \int f \, d\nu \tag{7.10}$$

for ν-almost all $(i_1, i_2, \dots)$. (See also Remark 7.38.) Showing below that $\log \circ \delta \circ \pi$ is an integrable function and taking $f := \log \circ \delta \circ \pi \circ \Theta - \log \circ \delta \circ \pi$ we infer

$$\lim_{n \to \infty} \frac{1}{n} \log \delta(\pi(\Theta^n(i_1, i_2, \dots)))$$

$$= \lim_{n \to \infty} \left(\frac{1}{n} \sum_{k=1}^{n} \left(\log \delta(\pi(\Theta^k(i_1, i_2, \dots))) - \log \delta(\pi(\Theta^{k-1}(i_1, i_2, \dots))) \right) \right.$$

$$\left. + \frac{1}{n} \log \delta(\pi(i_1, i_2, \dots)) \right) = \int (\log \circ \delta \circ \pi \circ \Theta - \log \circ \delta \circ \pi) \, d\nu = 0$$

for ν-almost all $(i_1, i_2, \dots) \in I^{\mathbb{N}}$, since ν is Θ-invariant. This proves equation (7.9).

The desired *integrability of the logarithmic distance function* is also of independent interest (cf. [Graf (1995)] for the Euclidean case):

Proposition 7.31. *Let $S_1, \dots, S_N$ be contracting similarities of the complete metric space (X, d) satisfying (SOSC) for the set O and let μ be the invariant normed measure with weights $p_1, \dots, p_N > 0$. Then we have*

$$\int |\log d(x, \partial O)| \, \mu(dx) < \infty.$$

Proof. Choose $(k_1, \ldots, k_m)$ as in the proof of Proposition 7.30, i.e.,
$S_{k_1} \circ \cdots \circ S_{k_m}(\overline{O}) \subset O$.

Denote

$$d(E, F) := \inf\{d(x, y) : x \in E, \ y \in F\} \text{ for closed } E, F \subset X,$$

$$\varrho := d(S_{k_1} \circ \cdots \circ S_{k_m}(\overline{O}), \partial O)\, r_{min},$$

$$a := r_{k_1} \cdots r_{k_m} r_{min} \qquad (0 < a < 1),$$

$$b := 1 - p_{k_1} \cdots p_{k_m} \qquad (0 < b < 1).$$

It suffices to show that

$$\mu\{x \in A : d(x, \partial O) < \varrho a^n\} \leq b^n, \quad n \in \mathbb{N}, \tag{7.11}$$

since in this case

$$\int |\log d(x, \partial O)|\, \mu(dx)$$

$$\leq \text{const} \sum_{n=1}^{\infty} \int_{\varrho a^{n+1} \leq d(x, \partial O) < \varrho a^n} |\log d(x, \partial O)|\, \mu(dx)$$

$$\leq \text{const} + \sum_{n=1}^{\infty} |\log(\varrho a^{n+1})|\, b^n < \infty.$$

Consider the Markov stop

$$M_n := \{(i_1, \ldots, i_\tau) : i_j \in I, \ r_{i_1} \cdots r_{i_\tau} < a^n \leq r_{i_1} \cdots r_{i_{\tau-1}}$$

$$(\leq r_{i_1} \cdots r_{i_\tau} r_{min}^{-1})\}$$

and denote

$$M_n' := \{(i_1, \ldots, i_\tau) \in M_n : d(S_{i_1} \circ \cdots \circ S_{i_\tau}(\overline{O}), \partial O) \leq \varrho a^n\}.$$

We next will show that

$$\sum_{\omega \in M_{n+1}'} \mu(A_\omega) \leq b \sum_{\omega \in M_n'} \mu(A_\omega), \tag{7.12}$$

where

$$E_\omega := S_{i_1} \circ \cdots \circ S_{i_\tau}(E), \text{ if } \omega = (i_1, \ldots, i_\tau) \text{ and } E \subset X.$$

For each $\omega \in M_{n+1}'$ there is a unique $\sigma \in M_n$ such that σ is a predecessor of ω, in symbols $\sigma \prec \omega$ (i.e., the sequence σ agrees with the first part of the sequence ω). Moreover, $\sigma \in M_n'$ because of $\overline{O}_\omega \subset \overline{O}_\sigma$ and therefore

$$d(\overline{O}_\sigma, \partial O) \leq d(\overline{O}_\omega, \partial O) \leq \varrho a^{n+1} < \varrho a^n.$$

Thus, the left-hand side of (7.12) may be split as follows:

$$\sum_{\omega \in M'_{n+1}} \mu(A_\omega) = \sum_{\sigma \in M'_n} \sum_{\substack{\omega \in M'_{n+1} \\ \sigma \prec \omega}} \mu(A_\omega).$$

Now for (7.12) it is enough to prove that

$$\sum_{\substack{\omega \in M'_{n+1} \\ \sigma \prec \omega}} \mu(A_\omega) \leq b\,\mu(A_\sigma), \quad \sigma \in M'_n. \tag{7.13}$$

To this aim we use the concatenation $\sigma * k = (i_1, \ldots, i_\tau, k_1, \ldots, k_m)$ for the above sequence k, the relation $r_{i_1} \cdots r_{i_\tau} r_{k_1} \cdots r_{k_m} \geq (a^n r_{min})(r_{min}^{-1}\, a) = a^{n+1}$ and that

$$\sigma * k \prec \omega \in M_{n+1} \quad \text{implies} \quad \omega \notin M'_{n+1},$$

since in this case $d(\overline{O}_\omega, \partial O) > \varrho a^{n+1}$. (The latter can be seen as follows:

$$d(\overline{O}_\omega, \partial O) \geq d(\overline{O}_{\sigma*k}, \partial O) \geq d(\overline{O}_{\sigma*k}, \partial(O_\sigma))$$
$$= r_{i_1} \cdots r_{i_\tau}\, d(S_{k_1} \circ \cdots \circ S_{k_m}(\overline{O}), \partial O)$$
$$\geq (a^n r_{min})(r_{min}^{-1}\varrho) = \varrho a^n > \varrho a^{n+1},$$

because $\sigma \in M_n$ and $d(S_{k_1} \circ \cdots \circ S_{k_m}(\overline{O}), \partial O) = r_{min}^{-1}\varrho$.)

Therefore the sum in (7.13) may be estimated by

$$\sum_{\substack{\omega \in M'_{n+1} \\ \sigma \prec \omega}} \mu(A_\omega) \leq \sum_{\substack{\omega \in M_{n+1} \\ \sigma \prec \omega}} \mu(A_\omega) - \sum_{\substack{\omega \in M'_{n+1} \\ \sigma * k \prec \omega}} \mu(A_\omega)$$

$$= \mu(A_\sigma) - \mu(A_{\sigma*k}) = \mu(A_\sigma) - p_{k_1} \cdots p_{k_m}\, \mu(A_\sigma)$$
$$= (1 - p_{k_1} \cdots p_{k_m})\mu(A_\sigma) = b\,\mu(A_\sigma).$$

In the first two equalities we have used Proposition 7.30 (ii) and (iii), respectively.

From (7.12) we infer by induction and 7.30 (ii)

$$\sum_{\omega \in M'_n} \mu(A_\omega) \leq b^n \sum_{\omega \in M'_0} \mu(A_\omega) \leq b^n. \tag{7.14}$$

Finally,

$$\{x \in A : d(x, \partial O) < \varrho\, a^n\} \subset \bigcup_{\omega \in M'_n} A_\omega,$$

since for such x there exists $\omega \in M_n$ with $x \in A_\omega \subset \overline{O}_\omega$, i.e., $d(\overline{O}_\omega, \partial O) < \varrho\, a^n$ which proves $\omega \in M'_n$. Hence,

$$\mu(\{x \in A : d(x, \partial O) < \varrho\, a^n\}) \leq \mu\left(\bigcup_{\omega \in M'_n} A_\omega \right) = \sum_{\omega \in M'_n} \mu(A_\omega)$$

by 7.30 (ii).

From this and (7.14) we get (7.11) and the proof is complete. $\qquad \square$

Exercise 7.5.

(i) The logarithmic distance function $|\log d(x, \partial O)|$ has moments of all orders with respect to the measure μ.

(ii) Prove that

$$\lim_{n \to \infty} \frac{1}{n} \log \delta(\pi(i_{n+1}, i_{n+2}, \dots)) = 0$$

at ν-almost all $(i_1, i_2, \dots)$ by means of the Borel-Cantelli lemma.

Summarizing the above properties we turn to the *main local dimension result for self-similar sets in metric spaces.*

Theorem 7.32. *Let $S_1, \dots, S_N$ be contracting similarities of a complete metric space with ratios $r_1, \dots, r_N$ satisfying the Strong Open Set Condition. A is the corresponding self-similar set and D given by $\sum_{i=1}^{N} r_i^D = 1$ is its similarity dimension. Let μ be the normalized self-similar measure with weights $p_i > 0$, i.e., $\mu = \sum_{i=1}^{N} p_i \, \mu \circ S_i^{-1}$. The we have the following :*

(i) $\underline{\dim}\mu(x) = \overline{\dim}\mu(x) = \frac{\sum_{i=1}^{N} p_i \log p_i}{\sum_{i=1}^{N} p_i \log r_i} =: \mathfrak{D}(p_1, \dots, p_N)$

at μ-almost all x, i.e., μ is dimension-regular and of exact Hausdorff and packing dimension $\mathfrak{D}(p_1, \dots, p_N)$.

(ii) $\mathfrak{D}(r_1^D, \dots r_N^D) = D = \dim_H A = \dim_P A = \dim_E A$, *provided the metric space is also separable.*

(iii) $\mathfrak{D}(p_1, \dots, p_N) < D$ *if* $(p_1, \dots, p_N) \neq (r_1^D, \dots, r_N^D)$.

Remark 7.33. Recall that in Theorem 7.19 we have shown under the additional assumption that A is a β-space the inequality $\mathcal{H}^D(A) > 0$ ($\mathcal{H}^D(A) < \infty$ was proved before for general contractions). By assertion (ii) the Strong Open Set Condition alone implies the weaker property $\dim_H A = D$. In distinction to the Euclidean case it can happen $\mathcal{H}^D(A) = 0$ (cf. [Schief (1996)]).

Proof of Theorem 7.32. Assertion (i) follows from the preceding considerations.

Putting in (i) $p_i := r_i^D$, $i = 1, \dots, N$, we obtain $\mathfrak{D}(r_1^D, \dots, r_N^D) = D$ and

$$\underline{\dim}\mu(x) = \overline{\dim}\mu(x) = D \quad \text{at } \mu\text{-a.a. } x.$$

Hence, $\dim_H \mu = \dim_P \mu = D$. From Theorem 7.10 we know that $\dim_H A \leq D$. Then the variational principle 6.7 (i) for Hausdorff dimension implies $\dim_H A = D$. (There we have used separability of the space.) Furthermore, the arguments in Section 7.3 for $\overline{\dim}_E A \leq D$ (see Theorem 7.17 together with Corollary 5.10) remain valid for arbitrary contractions in a complete metric space. Since $\dim_H A \leq \dim_P A \leq \overline{\dim}_E A$ the proof of (ii) is completed.

(iii) may be treated as an extremal problem for the function $\mathfrak{D}$: $(0, \infty)^N \to \mathbb{R}$ given by

$$\mathfrak{D}(p_1, \ldots, p_N) := \frac{\sum_{n=1}^{N} p_i \log p_i}{\sum_{i=1}^{n} p_i \log r_i}$$

with boundary condition $\sum_{i=1}^{N} p_i = 1$. Standard arguments (Lagrange method) lead to the maximal value D with unique extremal point $(p_1, \ldots, p_N) = (r_1^D, \ldots, r_N^D)$. $\qquad\square$

7.7 Average densities of self-similar measures

We now consider the case of similarities $S_1, \ldots, S_N$ satisfying SOSC, $(p_1, \ldots, p_N) = (r_1^D, \ldots, r_N^D)$ and the self-similar measure μ. In view of Theorem 7.20,

$$\mu = \nu \circ \pi^{-1} = \mathcal{H}^D(A)^{-1} \mathcal{H}^D \llcorner A$$

for the corresponding self-similar set A. Our aim is to calculate the average D-density of μ in the sense of Definition 3.4. As in the previous section we will use the ergodic dynamical system $(I^{\mathbb{N}}, \nu, \Theta)$ for the left shift operator $\Theta(i_1, i_2, \ldots) = (i_2, i_3, \ldots)$. Recall that for μ-a.a. $y \in A$ the coding sequence with $y = (i_1, i_2, \ldots)$ is uniquely determined. Therefore we also use the notations $(y_1, y_2, \ldots) := (i_1, i_2, \ldots)$ and $y|n := (y_1, \ldots, y_n)$ for such y and $n \in \mathbb{N}$. Then we get the following.

Theorem 7.34. *Suppose that the self-similar set A in $\mathbb{R}^d$ with IFS $S_1, \ldots, S_N$, contraction ratios $r_1, \ldots, r_N$, and Hausdorff dimension D satisfies the Strong Open Set Condition w.r.t. O. Then for μ-a.a. $x \in A$ the average density*

$$\mathcal{D}_a^D \mu(x) := \lim_{\delta \to 0} \frac{1}{|\ln \delta|} \int_\delta^1 \frac{\mu(B(x, \varepsilon))}{(2\varepsilon)^D} \varepsilon^{-1} d\varepsilon \qquad (7.15)$$

exists and equals the constant

$$\left(\sum_{j=1}^{N} r_j^D |\ln r_j| \right)^{-1} \int \int_{d(y, \partial(S_{y_1}(O)))}^{d(y, \partial O)} \frac{\mu(B(y, \varepsilon))}{(2\varepsilon)^D} \varepsilon^{-1} d\varepsilon \, \mu(dy). \qquad (7.16)$$

Proof. In the sequel we will always consider only μ-almost all points of $x \in A$, where the above identification with $(x_1, x_2, \ldots)$ is possible. Then define the mapping

$$T(x) := (S_j)^{-1}(x) \text{ if } x \in S_j(F), \ j = 1, \ldots N.$$

In the above notations this means that

$$T(x) = \Theta(x_1, x_2, \ldots).$$

Since $\mu = \nu \circ \pi^{-1}$, (A, T, μ) is also an ergodic dynamical system (cf. Remark 7.38 below).

Recall that $S_{x|i} = S_{x_1} \circ \ldots \circ S_{x_i}$ and $r_{x|i} = r_{x_1} \cdots r_{x_i}$. If $\varepsilon < d(x, \partial S_{x|i}(O))$, then

$$\mu(B(x, \varepsilon)) = r_{x|i}^D \, \mu(B(T^i x, r_{x|i}^{-1} \varepsilon)), \ i \in \mathbb{N}.$$

(To see this note that in this case $A \cap B(x, \varepsilon) = S_{x|i}(A) \cap B(x, \varepsilon)$ and therefore

$$\mathcal{H}^D(A) \, \mu(B(x, \varepsilon)) = \mathcal{H}^D(A \cap B(x, \varepsilon)) = \mathcal{H}^D(S_{x|i}(A) \cap B(x, \varepsilon))$$
$$= \mathcal{H}^D(S_{x|i}(A) \cap_{x|i} (B(T^i x, r_{x|i}^{-1} \varepsilon))) = r_{x|i}^D \, \mathcal{H}^D(A \cap B(T^i x, r_{x|i}^{-1} \varepsilon))$$
$$= \mathcal{H}^D(A) \, r_{x|i}^D \, \mu(B(T^i x, r_{x|i}^{-1} \varepsilon)),$$

since the $S_{x|i}$ are similarities.)

Now we will verify the limit

$$\lim_{\delta \to 0} \frac{1}{|\ln \delta|} \int_\delta^1 \frac{\mu(B(x, \varepsilon))}{(2\varepsilon)^D} \varepsilon^{-1} d\varepsilon = \lim_{\delta \to 0} \frac{1}{|\ln \delta|} \int_\delta^{d(x, \partial O)} \frac{\mu(B(x, \varepsilon))}{(2\varepsilon)^D} \varepsilon^{-1} d\varepsilon$$

$$= \lim_{\delta \to 0} \frac{n(x, \delta)}{|\ln \delta|} \frac{1}{n(x, \delta)} \left(\sum_{i=0}^{n(x,\delta)-1} \int_{d(x, \partial(S_{x|(i+1)}(O)))}^{d(x, \partial(S_{x|i}(O)))} \frac{\mu(B(x, \varepsilon))}{(2\varepsilon)^D} \varepsilon^{-1} d\varepsilon \right.$$

$$\left. + \int_\delta^{d(x, \partial(S_{x|n(x,\delta)}(O)))} \frac{\mu(B(x, \varepsilon))}{(2\varepsilon)^D} \varepsilon^{-1} d\varepsilon \right),$$

where

$$n(x, \delta) := \max\{n \in \mathbb{N} : d(x, \partial(S_{x|n}(O))) \geq \delta\}.$$

The first equality, provided the limit exists, and the measurability properties follow from the boundedness of the integrand $\mu(B(x, \varepsilon))/(2\varepsilon)^D$ (see Theorem 7.17) and Proposition 3.5, resp. By the above relationship the integrand in the ith integral in the last sum may be replaced by

$$r_{x|i}^D (2\varepsilon)^{-D} \, \mu(B(T^i x, r_{x|i}^{-1} \varepsilon)) \, \varepsilon^{-1}.$$

For the integral bounds we use

$$d(x, \partial(S_{x|i}(O))) = r_{x|i}\, d(T^i x, \partial O)\,,$$
$$d(x, \partial(S_{x|(i+1)}(O))) = r_{x|i}\, d(T^i x, \partial((T^i x)_1(O)))\,.$$

Substituting then under the integral $r_{x|i}^{-1}\,\varepsilon$ by ε we obtain the expression

$$\int_{d(T^i x, \partial(S_{(T^i x)_1}(O)))}^{d(T^i x, \partial O)} \frac{\mu(B(T^i x, \varepsilon))}{(2\varepsilon)^D}\, \varepsilon^{-1} d\varepsilon\,.$$

Therefore it suffices to show that for μ-a.a. $x \in A$ the following integrals and limit relationships exist:

$$\lim_{n\to\infty} \frac{1}{n} \sum_{i=1}^{n} \int_{d(T^i x, \partial(S_{(T^i x)_1}(O)))}^{d(T^i x, \partial O)} \frac{\mu(B(T^i x, \varepsilon))}{(2\varepsilon)^D}\, \varepsilon^{-1} d\varepsilon$$
$$= \int \int_{d(y, \partial(S_{y_1}(O)))}^{d(y, \partial O)} \frac{\mu(B(y, \varepsilon))}{(2\varepsilon)^D}\, \varepsilon^{-1} d\varepsilon\, \mu(dy)\,, \qquad (7.17)$$

$$\lim_{n\to\infty} \frac{1}{n} \int_{d(T^n x, \partial(S_{(T^n x)_1}(O)))}^{d(T^n x, \partial O)} \frac{\mu(B(T^n x, \varepsilon))}{(2\varepsilon)^D}\, \varepsilon^{-1} d\varepsilon = 0\,, \qquad (7.18)$$

(note that under the above conditions $d(T^n x, \partial((T^n x)_1(O))) < \delta$), and

$$\lim_{\delta\to 0} \frac{|\ln \delta|}{n(x, \delta)} = \sum_{j=1}^{N} r_j^D\, |\ln r_j|\,. \qquad (7.19)$$

Equation (7.17) follows from Birkhoff's ergodic theorem applied to the ergodic dynamical system (A, μ, T). Here the integrability condition follows from the fact that $\varepsilon^{-D}\mu(B(x, \varepsilon))$ is bounded by a constant (recall Theorem 3.5) together with

$$\int |\ln d(y, \partial O)|\, \mu(dy) < \infty \quad \text{(cf. Proposition 7.31)}\,.$$

Taking into regard that $a_n = \sum_{i=1}^{n} a_i - \sum_{i=1}^{n-1} a_i$ for any real sequence, (7.18) is a consequence.

In order to use these arguments for (7.19), too, note that for $\delta(x, n) := d(x, \partial(_{x|n}(O)))$ we get

$$\lim_{\delta\to 0} \frac{|\ln \delta|}{n(x, \delta)} = \lim_{n\to\infty} \frac{|\ln \delta(x, n)|}{n}\,,$$

provided the last limit exists. Since

$$\delta(x, n) = r_{x|n}\, d(T^n x, \partial O) = \prod_{i=1}^{n} r_{x_i}\, d(T^n x, \partial O)$$

and $x_i = (T^i x)_1$, $i \in \mathbb{N}$, Birkhoff's ergodic theorem (or more specific, the law of large numbers) implies for μ-a.a. $x \in A$,

$$\lim_{n \to \infty} \frac{1}{n} |\ln \prod_{i=1}^{n} r_{x_i}| = \lim_{n \to \infty} \frac{1}{n} \sum_{i=1}^{n} |\ln r_{(T^i x)_1}| = \int |\ln r_{y_1}| \, \mu(dy) = \sum_{j=1}^{N} |\ln r_j| \, r_j^D \,,$$

as well as

$$\lim_{n \to \infty} \frac{1}{n} |\ln d(T^n x, \partial O)| = 0 \,,$$

This shows (7.19) and thus, the proof of the theorem is completed. □

Remark 7.35. For the special case of Cantor type sets on the real line average densities were first considered in [Bedford and Fisher (1992)]. In [Patzschke and Zähle, M. (1994)] the case of self-affine measures was treated by means of Palm measures. The above approach can be found in [Graf (1995)]. The case of self-conformal measures was studied in [Zähle, M. (2001)]. In particular, an expression for the above average densities in the potential theoretic language of Chapter 4 was derived: Under the conditions of Theorem 7.34 we have for μ-a.a. x,

$$\mathcal{D}_a^D \mu(x) = \left(D \sum_{j=1}^{N} r_j^D |\ln r_j| \right)^{-1} \sum_{i \neq j} \int_{S_i(A)} \int_{S_j(A)} \frac{1}{|y - z|^D} \, \mu(dy) \, \mu(dz) \,.$$

(See also [Falconer (1997)] for Cantor-type sets.)

7.8   An ergodic theorem and applications in computer graphics

Recall that for a sequence $(\xi_1, \xi_2 \dots)$ of random indexes with distribution ν as before, the random variable

$$x_{\xi_1, \xi_2 \dots} = \pi(\xi_1, \xi_2, \dots) = \lim_{n \to \infty} \varphi_{\xi_1} \circ \cdots \circ \varphi_{\xi_n}(x_0)$$

possesses the distribution μ. From Theorem 7.19 we know that

$$\mu = \mathcal{H}^D(A)^{-1} \mathcal{H}^D \llcorner A$$

provided $\varphi_1, \dots, \varphi_N$ are similarities satisfying (SOSC), the self-similar set A is a β-space and $p_i = r_i^D$, $i = 1, \dots, N$. So in this case μ may be interpreted as *uniform distribution* on A. Moreover, according to Theorem 7.32 for any other choice of the probabilities $p_1, \dots, p_N$ the measure μ (with compact support A) is different from the uniform distribution.

Note that the order of application of the mappings $\varphi_{\xi_1}, \ldots, \varphi_{\xi_n}$ in the above limit makes related computer simulations rather extensive. The next *ergodic theorem* provides a much simpler method for generating μ, since there the mappings are applied in the natural order:

Theorem 7.36. *Let* $\varphi_1, \ldots, \varphi_N$ *be contractions of the complete metric space* (X, d) *and* μ *be the corresponding invariant probability measure on* $[X, \mathfrak{B}(X)]$ *with weights* $p_1, \ldots, p_N$. *Then for any continuous real-valued function* f *on* X *we have*

$$\lim_{n \to \infty} \frac{1}{n} \sum_{k=1}^{n} f(\varphi_{i_k} \circ \cdots \circ \varphi_{i_1}(x_0)) = \int f \, d\mu \,,$$

for any starting point $x_0 \in X$ *and* $\nu-$*almost all* $(i_1, i_2, \ldots) \in I^{\mathbb{N}}$.

Remark 7.37.

(i) The continuous function f may be replaced by the indicator function 1_B of a bounded open set B, if μ has no mass at the boundary of B. (Use that 1_B can be approximated from above and from below by continuous functions, say f_n and g_n, respectively, and that $\lim_{n \to \infty} (\int f_n d\mu - \int g_n d\mu) = 0$.)

(ii) This property may be used for *computer simulations* as follows. The space X is identified with the screen which is divided into boxes (pixels) B which are assumed to be of zero boundary mass with respect to μ. Then for the random variables $\xi_1, \xi_2, \ldots$ as before the value

$$\frac{1}{n} \sum_{k=1}^{n} 1_B(\varphi_{\xi_k} \circ \cdots \circ \varphi_{\xi_1}(x_0)) =: \frac{N_n(B)}{n}$$

is the *relative frequency* of meeting the box B in n random iterations of the mappings $\varphi_1, \ldots, \varphi_N$ with index probabilities $p_1, \ldots, p_N$. For large n it agrees approximately with $\mu(B)$. For different B this relative frequencies may be visualized by means of a color spectrum.

Before we turn to the proof of the ergodic theorem for iterated function systems we will briefly recall *Birkhoff's ergodic theorem* in a general context.

Remark 7.38. Let $[Y, \mathcal{Y}]$ be a measurable space and $T : Y \to Y$ a $\mathcal{Y}$-measurable mapping. Then $(Y, \{T^n\}_{N=1}^{\infty})$ is a *discrete dynamical system*.

A *T-invariant probability measure* m on $[Y, \mathcal{Y}]$ (i.e., $m = m \circ T^{-1}$) is said to be *ergodic* if any T-invariant set $E \in \mathcal{Y}$ (i.e., $T^{-1}(E) = E$) has m-measure either zero or one.

The Birkhoff ergodic theorem states that for such measures m and any m-integrable real-valued function g on Y we have

$$\lim_{n \to \infty} \frac{1}{n} \sum_{k=1}^{n} g(T^k y) = \int g\, dm$$

at m-almost all y. (See, e.g., [Walters (1982)].)

Proof of Theorem 7.36. In order to apply Birkhoff's ergodic theorem we set $Y := X \times I^{\mathbb{N}}$, $\mathcal{Y} := \mathfrak{B}(X) \otimes \mathfrak{B}(I^{\mathbb{N}})$, $m := \mu \times \nu = \nu \circ \pi^{-1} \otimes \nu$ and define $T : X \times I^{\mathbb{N}} \to X \times I^{\mathbb{N}}$ by

$$T(x, (i_1, i_2, \dots)) = (\varphi_{i_1}(x), \Theta(i_1, i_2, \dots)) = (\varphi_{i_1}(x), (i_2, i_3, \dots)).$$

We have to check that m is T-invariant and ergodic.

Denote $\bar{i} := (i_1, i_2, \dots) \in I^{\mathbb{N}}$ and let $E \in \mathcal{Y}$. Then we obtain, because of the structure of ν,

$$
\begin{aligned}
m(T^{-1}(E)) &= \int 1_E(Ty)\, m(dy) = \int \int 1_E(\varphi_{i_1}(x), \Theta\bar{i})\, \mu(dx)\nu(d\bar{i}) \\
&= \int \int 1_E(\varphi_{i_1}(\pi(\bar{j})), \Theta\bar{i})\, \nu(d\bar{j})\nu(d\bar{i}) \\
&= \int \int 1_E(\pi(i_1\bar{j}), \Theta\bar{i})\, \nu(d\bar{j})\nu(d\bar{i}) \\
&= \int \int 1_E(\pi(\bar{l}), \bar{k})\, \nu(d\bar{l})\nu(d\bar{k}) \\
&= \int \int 1_E(x, \bar{k})\, \mu(dx)\nu(d\bar{k}) = m(E)\,,
\end{aligned}
$$

i.e., T-invariance of μ. Let now $E \in \mathcal{Y}$ be a T-invariant set and choose (for the identity id),

$$E_k := (\pi^{-1} \times \mathrm{id}(\{(\bar{j}, \bar{i}) \in I^{\mathbb{N}} \times I^{\mathbb{N}} : (j_1, \dots, j_k, i_1, \dots, i_k) \in D_k\})$$

with $D_k \in I^{2k}$ such that $m(E \triangle E_k) < \varepsilon(k) \to 0$, as $k \to \infty$. (This is possible, since the ring of sets of the form E_k generates the σ-algebra $\mathcal{Y}$.)

By the invariance of m it follows that

$$m(T^{-n}(E) \triangle T^{-n}(E_k)) = m(E \triangle E_k) < \varepsilon(k)$$

and by the invariance of E,

$$m(E) = m(T^{-n}(E) \cap E) = m(T^{-n}(E_k) \cap E_k) + \alpha(k)\,,$$

where $\alpha(k) < 2\varepsilon(k)$. For $n > 2k$ we obtain

$$T^{-n}(E_k) = (\pi^{-1} \times \mathrm{id})(\{(\bar{j}, \bar{i}) : (j_n, \dots, j_{n-k}, i_{n+1}, \dots, i_{n+k}) \in D_k\})$$

and hence,

$$m(T^{-n}(E_k) \cap E_k) = \nu \otimes \nu(\{(\bar{j}, \bar{i}) : (j_n, \dots, j_{n-k}, i_{n+1}, \dots, i_{n+k}) \in D_k,$$
$$(j_1, \dots, j_k, i_1, \dots, i_k) \in D_k\})$$
$$= \nu \otimes \nu(\{(\bar{j}, \bar{i}) : (j_1, \dots, j_k, i_1, \dots, i_k) \in D_k\} = m(E_k)^2 = m(E)^2 + \beta(k),$$

where $\beta(k) < 3\varepsilon(k)$. Therefore

$$m(E) = m(E)^2 + \gamma(k) \quad \text{with} \quad \lim_{k \to \infty} \gamma(k) = 0.$$

Thus, $m(E) = m(E)^2$, which implies $m(E) \in \{0, 1\}$, i.e., ergodicity of m.

For any continuous function f on X we now define $g : X \times I^{\mathbb{N}} \to \mathbb{R}$ by

$$g(x, \bar{i}) := f(\varphi_{i_1}(x)).$$

Then g is m-integrable and

$$\int g\, dm = \int \int f(\varphi_{i_1}(x))\, \mu(dx)\nu(d\bar{i})$$
$$= \sum_{i=12}^{N} p_i \int f(\varphi_i(x))\, \mu(dx) = \int f\, d\mu,$$

in view of the invariance property of μ.

Birkhoff's ergodic theorem now leads to

$$\lim_{n \to \infty} \frac{1}{n} \sum_{k=1}^{n} f(\varphi_{i_k} \circ \cdots \circ \varphi_{i_1}(x)) = \lim_{n \to \infty} \frac{1}{n} \sum_{k=1}^{n} g(T^k(x, \bar{i}))$$
$$= \int g\, dm = \int f\, d\mu,$$

for m-almost all $(x, \bar{i}) \in X \times I^{\mathbb{N}}$.

Finally, we will show that on the left-hand side x may be replaced by any starting point x_0:

$$\lim_{n \to \infty} \left| \frac{1}{n} \sum_{k=1}^{n} f(\varphi_{i_k} \circ \cdots \circ \varphi_{i_1}(x_0)) - \frac{1}{n} \sum_{k=1}^{n} f(\varphi_{i_k} \circ \cdots \circ \varphi_{i_1}(x)) \right|$$
$$\leq \lim_{n \to \infty} \frac{1}{n} \sum_{k=1}^{n} \left| f(\varphi_{i_k} \circ \cdots \circ \varphi_{i_1}(x_0)) - f(\varphi_{i_k} \circ \cdots \circ \varphi_{i_1}(x)) \right|$$
$$\leq \lim_{n \to \infty} \left| f(\varphi_{i_n} \circ \cdots \circ \varphi_{i_1}(x_0)) - f(\varphi_{i_n} \circ \cdots \circ \varphi_{i_1}(x)) \right| = 0,$$

since

$$d(\varphi_{i_n} \circ \cdots \circ \varphi_{i_1}(x_0), \varphi_{i_n} \circ \cdots \circ \varphi_{i_1}(x)) \leq r_{max}^n d(x_0, x)$$

and f is continuous. $\qquad \square$

Chapter 8

An Example from the Theory of Dynamical Systems

Fractal limit sets of certain dynamical systems may be interpreted as special cases of the invariant sets from the previous chapter. These and other models have been considered in the literature on mathematics and applied sciences in various contexts, in particular, in the thermodynamic formalism. For some mathematical background see, e.g., [Pesin (1997)], [Mauldin and Urbanski (2003)], [Pesin and Climenhaga (2009)], [Przytycki and Urbanski (2010)], [Barreira (2011)] and the references therein. We will demonstrate some basic approach on the simple model generated by the *logistic map* $f_\lambda : \mathbb{R} \to \mathbb{R}$ with parameter $\lambda > 0$ given by

$$f_\lambda(x) := \lambda x(1 - x),$$

which has been considered in various papers with different applications. (Cf. [Falconer (1990)] and the references therein.) A possible interpretation is the development of a population (see, e.g. the logistic map in Wikipedia for more details). The mathematical model behind became popular by the paper [May (2003)], where the diversity of the related periodic and chaotic dynamics was studied.

The parameter λ plays an essential role, when considering the orbits arising from iterated application of the mapping f_λ. We will briefly describe the behavior, first for starting points $x \in [0, 1]$ and different λ.

- For $\lambda = 2$ is illustrated in Figure 8.1 below. Here $x = 0$ is an unstable fixed point and $x = \frac{1}{2}$ a stable one.
- For general λ we use the derivative

$$f'_\lambda(x) = \lambda(1 - 2x) \quad \text{and} \quad f'_\lambda(0) = \lambda$$

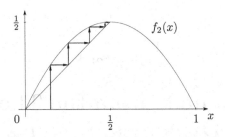

Fig. 8.1   The arrows show he movement of a starting point towards the stable fixed point.

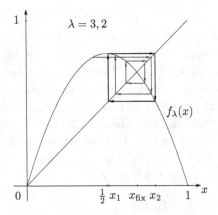

Fig. 8.2   Illustration for the case $\lambda = 3, 2$.

in order to conclude that the fixed point $x = 0$ is stable, if $\lambda < 1$, and unstable, if $\lambda > 1$. The second fixed point of f_λ is given by

$$x_\lambda = 1 - \frac{1}{\lambda}, \quad \text{and} \quad |f'_\lambda(x_\lambda)| = |\lambda - 2|,$$

so that x_λ is stable, if $1 < \lambda < 3$, and unstable, if $\lambda < 1$ or $\lambda > 3$.

- For λ slightly greater than 3 there arises a stable orbit of period 2 (see Figure 8.2).
- Here x_λ is an unstable fixed point, but there are two points x_1, x_2 such that $f_\lambda(x_1) = x_2$ and $f_\lambda(x_2) = x_1$, i.e., x_1 and x_2 are fixed points of $f_\lambda^2 = f_\lambda \circ f_\lambda$.
- x_1, x_2 are stable for f_λ^2, since $|(f_\lambda^2)'| < 1$ at these points. All orbits converge to the orbits x_1 or x_2. The attractor (x_1, x_2) is also called *stable orbit of f_λ of period* 2. (For $\lambda = 3$ the points x_1 and x_2 coincide.)
- When further increasing λ the values of $|(f_\lambda^2)'|$ at x_1 and x_2 converge to 1, and at a certain λ these points become unstable for f_λ^2. It arises a stable orbit of period 4, then of periods $8, 16, \ldots$.

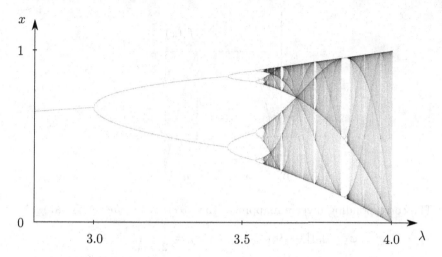

Fig. 8.3 The attractor corresponding to a fixed $2.8 \leq \lambda \leq 4$ is shown on the corresponding vertical line.

- The lengths of the intervals between the corresponding λ are decreasing for the sequence of doubling periods, and there exists a limit value $\lambda_\infty \approx 3.57$.
- More visual information can be obtained from the *bifurcation diagram* for the logistic map, where in dependence on λ several hundred of iterations of f_λ are plotted (sufficiently many primary iterations are omitted) (see Figure 8.3).
- For $\lambda_\infty < \lambda < 4$ there exist stable periodic orbits as well as unstable orbits and chaos.
- For $\lambda > 4$ the dynamics is always chaotic, which means that the orbits essentially depend on the starting point.

Our main aim is now to give a strong mathematical description of the last fact and to establish relationships to Chapter 7 for λ slightly greater than 4. We first suppose that $\lambda > 4$.

The mapping f_λ is not injective, but $f_\lambda^{-1}([0, 1])$ can be divided into two sub-intervals, where f_λ is injective: Let

$$a := \frac{1}{2} - \sqrt{\frac{1}{4} - \frac{1}{\lambda}}, \quad 1 - a = \frac{1}{2} + \sqrt{\frac{1}{4} - \frac{1}{\lambda}}.$$

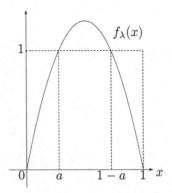

The corresponding inverse mappings (the *inverse branches* of f_λ) are

$$\varphi_1 : [0,1] \to [0,a], \qquad \varphi_1(x) := \frac{1}{2} - \sqrt{\frac{1}{4} - \frac{x}{\lambda}}$$

$$\varphi_2 : [0,1] \to [1-a,1], \quad \varphi_2(x) := \frac{1}{2} + \sqrt{\frac{1}{4} - \frac{x}{\lambda}}.$$

(For brevity we omit the dependence of λ in the notation.) Then we get for $x, y \in [0,1]$ and $i = 1, 2$,

$$|\varphi_i'(x)| = \frac{1}{2\lambda} \left(\frac{1}{4} - \frac{x}{\lambda} \right)^{-1/2}, \text{ and } |\varphi_i(x) - \varphi_i(y)| = |\varphi_i'(\theta)||x - y|$$

for some θ between x and y by the mean value theorem. Therefore

$$\frac{1}{\lambda}|x - y| \le |\varphi_i(x) - \varphi_i(y)| \le \frac{1}{2\lambda} \left(\frac{1}{4} - \frac{1}{\lambda} \right)^{-1/2}, \tag{8.1}$$

and φ_i is contractive if and only if

$$\frac{1}{2\lambda} \left(\frac{1}{4} - \frac{1}{\lambda} \right)^{-1/2} < 1, \quad \text{i.e.,} \quad \lambda > 2 + \sqrt{5}.$$

For such λ we now consider (φ_1, φ_2) as IFS in the sense of Section 7.1. Let here F_λ denote the corresponding attractor A, i.e., the set with

$$\varphi(F_\lambda) = F_\lambda \tag{8.2}$$

(for $\varphi(K) := \varphi_1(K) \cup \varphi_2(K)$, $K \subset [0,1]$, where the unions here are disjoint). From this we infer that

$$f_\lambda(F_\lambda) = F_\lambda, \tag{8.3}$$

since $f_\lambda(\varphi(F_\lambda)) = f_\lambda(\varphi_1(F_\lambda)) \cup f_\lambda(\varphi_2(F_\lambda))$ and by construction $f_\lambda(\varphi_i(F_\lambda)) = F_\lambda$ for $i = 1, 2$.

Next we will show that F_λ is a *repeller* for f_λ. This means that all points from a certain neighborhood of F_λ, which do not belong to F_λ, under iterated application of f_λ are moving arbitrarily far away. Recall that $f_\lambda(x) = \lambda x(1 - x)$. Then we get the following:

(i) $x < 0$: $f_\lambda^n(x) \to -\infty$ as $n \to \infty$.
(ii) $x > 1$: $f_\lambda^n(x) \to -\infty$ as $n \to \infty$.
(iii) $x \in [0, 1] \setminus F_\lambda$: $f_\lambda^n(x) \to -\infty$ as $n \to \infty$.

(i) and (ii) are easy to see. For (iii) note that for the case $x \in [0, 1] \setminus F_\lambda$ there exists some k such that $\varphi^k(x) \notin [0, 1]$ (Otherwise $x \in \bigcap_{k=1}^\infty \varphi^k(]0, 1]) = F_\lambda$.) Consequently, $f_\lambda^k(x) \notin [0, 1]$. (Otherwise $x \in \varphi^k(f_\lambda^k(x)) \subset \varphi^k([0, 1])$.) Then (i) and (ii) lead to (iii).

Chaotic behavior of the dynamics on the invariant set F_λ. This means that there exists a constant $c > 0$ such that for arbitrarily close starting points $x, y \in F_\lambda$ their orbit points after a certain iteration of f_λ have distance greater than c. Recall from Section 7.2 that

$$F_\lambda = \bigcup_{(i_1, i_2, \ldots) \in \{1,2\}^{\mathbb{N}}} \bigcap_{n=1}^\infty \varphi_{i_1} \circ \cdots \circ \varphi_{i_n}([0.1])$$

and $\bigcap_{n=1}^\infty \varphi_{i_1} \circ \cdots \circ \varphi_{i_n}([0, 1]) = \pi(i_1, i_2, \ldots)$ for the projection mapping π.

Let $x = \pi(i_1, i_2, \ldots)$ and $y = \pi(j_1, j_2, \ldots)$ be two different starting points from F_λ for the iterations of f_λ and $k := \min\{n : i_n \neq j_n\}$. We may assume that $i_k = 1$ and $j_k = 2$. Using again that the sets $\varphi_{i_1} \circ \cdots \circ \varphi_{i_n}([0, 1])$ are decreasing in n we obtain

$$f_\lambda^{k-1}(x) = f_\lambda^{k-1}\left(\bigcap_{n=1}^\infty \varphi_{i_1} \circ \cdots \circ \varphi_{i_n}([0, 1])\right) = f_\lambda^{k-1}\left(\bigcap_{n=k}^\infty \varphi_{i_1} \circ \cdots \circ \varphi_{i_n}([0, 1])\right)$$

$$= \bigcap_{n=k}^\infty f_\lambda^{k-1} \circ \varphi_{i_1} \circ \cdots \circ \varphi_{i_{k-1}} \circ \varphi_{i_k} \circ \cdots \circ \varphi_{i_n}([0, 1])$$

$$= \bigcap_{n=k}^\infty \varphi_{i_k} \circ \cdots \circ \varphi_{i_n}([0, 1]) \in \varphi_1([0, 1]),$$

since $f_\lambda^{k-1} \circ \varphi_{i_1} \circ \cdots \circ \varphi_{i_{k-1}}$ is the identity and $i_k = 1$. Hence, $f_\lambda^{k-1}(x) \in \varphi_1([0, 1])$. Similarly we get $f_\lambda^{k-1}(y) \in \varphi_2([0, 1])$. The distance between the intervals $\varphi_1([0, 1])$ and $\varphi_2([0, 1])$ is equal to

$$1 - 2a = 2\sqrt{\frac{1}{4} - \frac{1}{\lambda}} =: c.$$

Therefore the distance between the $(k-1)th$ iterates of the points x and y is at least c, which proves the chaotic behavior. (Note that k depends on the points.)

Dimension estimates for F_λ. Here we will use again some methods from the previous chapters. In view of the right-hand side of (8.1) the mappings φ_1 and φ_2 possess the contraction factors $r_1 = r_2 = \frac{1}{2\lambda}(\frac{1}{4} - \frac{1}{\lambda})$. Let $\overline{D} = \overline{D}_\lambda$ be determined by

$$r_1^{\overline{D}} + r_2^{\overline{D}} = 1, \quad \text{i.e.}, \quad \overline{D}_\lambda = \frac{\log 2}{\log \lambda + \frac{1}{2}\log(1 - \frac{4}{\lambda})} \tag{8.4}$$

and consider the product measure $\nu = \otimes_{i=1}^{\infty}\nu_i$ on $\{1,2\}^{\mathbb{N}}$ with $\nu_i(\{1\}) := r_1^{\overline{D}}$ and $\nu_i(\{2\}) := r_2^{\overline{D}}$ and the image measure $\mu := \nu \circ \pi^{-1}$. Then we get as in the case of similarities (see proof of Theorem 7.14) that for $x \in F_\lambda$ and $0 < r < 1$,

$$\frac{\mu(B(x,r))}{r^{\overline{D}}} \geq a > 0$$

for the constant a as above. Therefore Corollary 5.10 yields an *upper estimate for the upper entropy dimension*:

$$\overline{\dim}_E F_\lambda \leq \overline{D}_\lambda. \tag{8.5}$$

(As in Example 5.20 this can be shown here also with more direct arguments.)

For lower estimates we will use the left hand side of (8.1), i.e., $\frac{1}{\lambda}|x-y| \leq |\varphi_i(x) - \varphi_i(y)|$, and set now $\varrho_1 = \varrho_2 = \frac{1}{2\lambda}$. Let $\underline{D} = \underline{D}_\lambda$ be determined by

$$\varrho_1^{\underline{D}} + \varrho_2^{\underline{D}} = 1, \quad \text{i.e.}, \quad \underline{D}_\lambda = \frac{\log 2}{\log \lambda}, \tag{8.6}$$

and consider now the product measure $\nu = \otimes_{i=1}^{\infty}\nu_i$ on $\{1,2\}^{\mathbb{N}}$ with $\nu_i(\{1\}) := \varrho_1^{\underline{D}}$ and $\nu_i(\{2\}) := \varrho_2^{\underline{D}}$ and the image measure $\mu := \nu \circ \pi^{-1}$. Using the separation property of the IFS on obtains similarly as in the case of similarities (Theorem 7.14)

$$\frac{\mu(B(x,r))}{r^{\underline{D}}} \leq b, \quad 0 < r < 1,$$

for some constant $b > 0$ (see also [Falconer (1990), Proposition 9.7]). Therefore the mass distribution principle 3.8 leads to

$$\underline{\dim}_H F_\lambda \geq \underline{D}_\lambda. \tag{8.7}$$

Note that for $\lambda \to \infty$ the dimensions $\dim_H F_\lambda$, $\underline{\dim}_E F_\lambda$ and $\overline{\dim}_E F_\lambda$ are all close to $\frac{\log 2}{\log \lambda}$. Since in this case

$$\varphi_1(x) = \frac{1}{2} - \sqrt{\frac{1}{4} - \frac{x}{\lambda}} \approx \frac{1}{\lambda}x \quad \text{and} \quad \varphi_2(x) = \frac{1}{2} + \sqrt{\frac{1}{4} - \frac{x}{\lambda}} \approx 1 - \frac{1}{\lambda}x,$$

this interpretation agrees with the previous results for similarities.

Remark. An analysis of the above arguments shows that the approach can be extended to higher dimensional spaces and more general expanding maps than the f_λ, provided they possess a finite number of certain inverse branches. However, there exist more direct methods, e.g. in the literature mentioned above, for dynamical systems with associated invariant fractal sets.

Chapter 9

Graphs of Functions and Stochastic Processes

Fractal properties of non-smooth (random) functions are of interest in various applications. In this chapter we will estimate (or determine) the dimensions of their graphs (in particular, that of fractional Brownian motion) using the results from Chapters 3–5.

9.1 Hölder continuous functions

Let $I \subset \mathbb{R}$ be any finite interval of length $|I|$. Recall that for a function $f : I \mapsto \mathbb{R}$ the *graph* of f is defined as

$$\text{graph}(f) := \{(x, f(x)) : x \in I\}.$$

Hölder continuity of such f allows an *upper estimate for the box dimension* of $\text{graph}(f)$:

Lemma 9.1. *If*

$$|f(x) - f(y)| \le \text{const} \, |x - y|^{\alpha}, \quad x, y \in I,$$

for some $\alpha \in (0, 1]$, then

$$\overline{\dim}_B \text{graph}(f) \le 2 - \alpha.$$

Proof. Recall from Section 5.3 that

$$\overline{\dim}_B \text{graph}(f) = \varlimsup_{\delta \to 0} \frac{\log N_{\delta}^B(\text{graph}(f))}{|\log \delta|},$$

where $N_{\delta}^B(\text{graph}(f))$ is the number of boxes of a lattice with lattice constant δ which intersect $\text{graph}(f)$. The number of such boxes in a column of the lattice does not exceed $\frac{\text{const} \, \delta^{\alpha}}{\delta} + 2$. Hence,

$$N_{\delta}^B(\text{graph}(f)) \le \frac{|I|}{\delta} \left(\frac{\text{const} \, \delta^{\alpha}}{\delta} + 2 \right) = \delta^{-(2-\alpha)}(\text{const} + 2\delta^{1-\alpha}).$$

This implies

$$\varlimsup_{\delta \to 0} \frac{\log N_\delta^B(\mathrm{graph}(f))}{|\log \delta|} \leq 2 - \alpha,$$

which shows the assertion.   □

We now consider an interesting example from the literature. Let $\alpha \in (0, 1]$, $b > 1$ and a Hölder continuous function $g : [0, \infty) \to \mathbb{R}$ of order $\beta > \alpha$ with period 1 be given. Then

$$W_g^\alpha(x) := \sum_{n=0}^{\infty} b^{-n\alpha} g(b^n x), \ x \geq 0, \tag{9.1}$$

is called *Weierstrass-type function* with parameters α, b and generating function g. For the version $g(x) = \cos(\pi x)$ (with period 2) and special parameters this corresponds to the classical case introduced by Weierstrass in 1872. Here we include the more general variant (cf. Figure 9.1)

$$W(x) := \sum_{n=0}^{\infty} a^n \sin(b^n 2\pi x), \ 0 < a < 1, \ b > 1, \ ab > 1.$$

In [Hardy (1916)] it was proved that for such parameters W^α is nowhere differentiable.

Lemma 9.2. *Under the above conditions W_g^α is Hölder continuous of order α.*

Proof. Let $0 < |x - y| < 1$, choose $m \in \mathbb{N}$ such that

$$b^{-(m+1)} \leq |x - y| < b^{-m} \tag{9.2}$$

and $\|g\| := \sup_{x \in [0,1]} |g(x)|$. Then we get

$$|W_g^\alpha(x) - W_g^\alpha(y)|$$

$$\leq \sum_{n=0}^{m} b^{-n\alpha} |g(b^n x) - g(b^n y)| + \sum_{n=m+1}^{\infty} b^{-n\alpha} |g(b^n x) - g(b^n y)|$$

$$\leq \sum_{n=0}^{m} b^{-n\alpha} \mathrm{const}\, b^{n\beta} |x - y|^\beta + 2\|g\| \sum_{n=m+1}^{\infty} b^{-n\alpha}$$

$$\leq \frac{b^{m(\beta - \alpha)}}{1 - b^{-(\beta - \alpha)}} \mathrm{const}|x - y|^{\beta - \alpha} |x - y|^\alpha + 2\|g\| \frac{b^{-(m+1)\alpha}}{1 - b^{-\alpha}}$$

$$\leq \mathrm{const}|x - y|^\alpha$$

for varying constants. In the last estimate we have used (9.2). This proves the assertion.   □

The last two lemmas lead to the following upper estimate of the upper box dimension.

Corollary 9.3.
$$\overline{\dim}_B(\text{graph}(W_g^\alpha)) \leq 2 - \alpha.$$
for any Weierstrass-type function as above on a finite interval.

Remark 9.4. In the classical case the equality
$$\dim_B(\text{graph}(W^\alpha)) = 2 - \alpha$$
was first shown in [Kaplan, Mallet-Peret, Yorke (1984)]. For b large enough a short and simple proof may be found in [Falconer (1990), 11.3].

It was conjectured for many years that at least in the classical case the number $2 - \alpha$ also agrees with the Hausdorff dimension of the graph. The most general version was obtained only recently: In [Shen (2018)] it is shown the following. For any C^2-function g with period 1 and any integer $b \geq 2$ there exists a constant $K = K(g, b) > 1$ such that if $\alpha \in (0, 1]$ and $1 < b^{1-\alpha} < K$ then
$$\dim_H(\text{graph}(W_g^\alpha)) = 2 - \alpha.$$
Moreover, complete references to preceding papers and used techniques from there are given.

The methods become easier when considering graphs of certain stochastic processes, which possess scaling properties and bounded one-dimensional densities. This will be shown in the next section. In particular, applications to random versions of the Weierstrass-type functions are possible. To this aim it is useful to study a variant of W_g^α used in [Mandelbrot (1982)] for applications in physics. Under the additional condition $g(0) = 0$,

$$M_g^\alpha(x) := \sum_{n=-\infty}^{\infty} b^{-n\alpha} g(b^n x), \quad x \geq 0, \tag{9.3}$$

is called *Weierstrass-Mandelbrot-type function* (cf. Figure 9.2 below). It is easy to see that it possesses the scaling property

$$M_g^\alpha(x) = b^{-\alpha} M_g^\alpha(bx). \tag{9.4}$$

We now decompose

$$M_g^\alpha = W_g^\alpha + h^\alpha,$$

where $h^\alpha(x) = \sum_{n=1}^{\infty} b^{n\alpha} g(b^{-n} x)$.

Exercise 9.1. h^α is Hölder continuous of the same order β as the generating function g.

In a more general situation we obtain the following dimension relationship:

Lemma 9.5. *If a function φ on $[a, b]$ is representable as*

$$\varphi = \psi + h$$

for a Lipschitz continuous function h, then

$$\dim_H(\mathrm{graph}(\varphi)) = \dim_H(\mathrm{graph}(\psi)).$$

Proof. Define the mapping $f : \mathrm{graph}(\psi) \to \mathrm{graph}(\varphi)$ by

$$f(x, \psi(x)) := (x, \psi(x) + h(x)).$$

Then we get from $|h(x) - h(y)| \le C|x - y|$ that

$$
\begin{aligned}
|f(x, \psi(x)) - f(y, \psi(y))| &= (|x - y|^2 + |\psi(x) - \psi(y) + h(x) - h(y)|^2)^{1/2} \\
&\le (|x - y|^2 + (|\psi(x) - \psi(y)| + C|x - y|)^2)^{1/2} \\
&=: S.
\end{aligned}
$$

If $|\psi(x) - \psi(y)| \le |x - y|$, then

$$S \le \mathrm{const}|x - y| \le \mathrm{const}(|x - y|^2 + |\psi(x) - \psi(y)|^2)^{1/2}.$$

If $|x - y| < |\psi(x) - \psi(y)|$, then

$$S \le \mathrm{const}|\psi(x) - \psi(y)| \le \mathrm{const}(|x - y|^2 + |\psi(x) - \psi(y)|^2)^{1/2}.$$

Hence, the mapping f is Lipschitz continuous. Therefore Proposition 2.11 implies

$$\dim_H(\mathrm{graph}(\varphi)) \le \dim_H(\mathrm{graph}(\psi)).$$

Finally, exchanging the roles of φ and ψ and replacing h by $-h$ we infer the opposite inequality

$$\dim_H(\mathrm{graph}(\psi)) \le \dim_H(\mathrm{graph}(\varphi)),$$

and thus, the assertion. $\qquad\qquad\qquad\qquad\qquad\qquad\qquad\qquad\qquad\square$

Corollary 9.6. *For the Weierstrass-type function W_g^α as in (9.1) with $g(0) = 0$ and $\beta = 1$ and the corresponding Weierstrass-Mandelbrot-type function as in (9.3) one has*

$$\dim_H(\mathrm{graph}(M_g^\alpha)) = \dim_H(\mathrm{graph}(W_g^\alpha)).$$

In Figures 9.1 and 9.2 below we illustrate the behavior of the graphs of both functions for an example with $b = 3/2$ and $\alpha = 1/2$, i.e., where the box dimension equals 3/2.

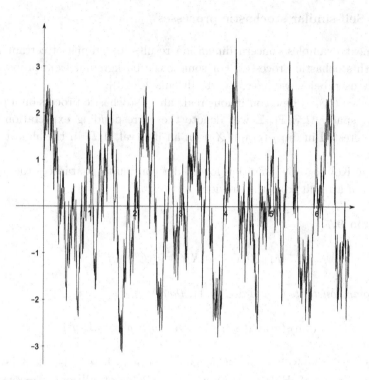

Fig. 9.1 The graph of a classical version of the Weierstrass function $W(x) = \sum_{n=0}^{\infty}(3/2)^{-(1/2)n}\sin((3/2)^n 2\pi x)$.

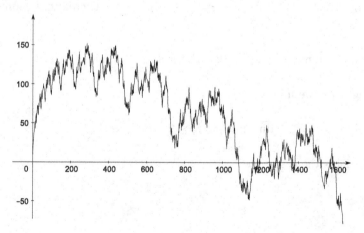

Fig. 9.2 The Weierstrass-Mandelbrot function $M(x) = \sum_{n=-\infty}^{\infty}(3/2)^{-(1/2)n}\sin((3/2)^n 2\pi x)$.

9.2　Self-similar stochastic processes

Our next examples concern dimension results for graphs of certain non-smooth stochastic processes. For some basic background from probability theory used below we refer, e.g., to [Klenke (2020)].

Let $(X(t))_{t \geq 0}$ be a continuous real valued stochastic process on a probability space $[\Omega, \mathfrak{F}, \mathbb{P}]$. $\mathbb{E}$ will denote the corresponding expectation. We are interested in $\dim_H(\text{graph}(X))$ on an interval $[0, T]$ in the almost sure sense.

The Kolmogorov condition for Hölder continuity guarantees the application of a result from the previous section:

Lemma 9.7.

$$If \quad \mathbb{E}(X(s) - X(t))^2 \leq C|s - t|^{2\alpha}$$

for some constants C and $\alpha \in (0, 1)$, then

$$\overline{\dim}_B(\text{graph}(X)) \leq 2 - \alpha \quad \text{with probability } 1.$$

Proof. By the Kolmogorov-Chentsov theorem there exists a subset A of the space of Hölder continuous functions of all orders less then α such that A is measurable with respect to the σ-algebra generated by $(X(t), t \geq 0)$ and $\mathbb{P}(X \in A) = 1$. Therefore Lemma 9.1 implies

$$\overline{\dim}_B \text{graph}(f) \leq 2 - \beta \quad \text{for all } f \in A \text{ and } \beta < \alpha.$$

Letting $\beta \uparrow \alpha$ we infer

$$\overline{\dim}_B \text{graph}(f) \leq 2 - \alpha, \quad f \in A,$$

which yields the assertion. □

A standard example is the *fractional Brownian motion* with Hurst parameter $\alpha \in (0,1)$. It is a continuous Gauss process wit $\mathbb{E}B^\alpha = 0$ and

$$\mathbb{E}(B^\alpha(s) - B^\alpha(t))^2 = |s - t|^{2\alpha}.$$

The covariance function is equal to

$$\mathbb{E}B^\alpha(s)B^\alpha(t) = \frac{1}{2}\left(\mathbb{E}B^\alpha(s)^2 + \mathbb{E}B^\alpha(t)^2 - \mathbb{E}(B^\alpha(s) - B^\alpha(t))^2\right)$$
$$= \frac{1}{2}\left(s^{2\alpha} + t^{2\alpha} - |s - t|^{2\alpha}\right).$$

For $\alpha = 1/2$ one obtains the classical one-dimensional Brownian motion, i.e., the *Wiener process*. Existence and properties of such processes, in particular those used below, are treated in many textbooks. More recent developments and related literature can be found, e.g., in [Tudor (2014)].

Definition 9.8. A process X as above possesses *stationary increments*, if the process

$$X_s(t) := X(t + s) - X(s), \ t \geq 0,$$

for any $s > 0$ has the same distribution as X.

Definition 9.9.

(1) Let $r > 0$ and $\alpha \in (0,1)$. Then a process X is said to be (r,α)-*scaling* or (r,α)-*self-similar*, if the process

$$X_{r,\alpha}(t) := r^{-\alpha}X(rt), \ t \geq 0,$$

has the same distribution as X

(2) X is said to be α-*scaling* or α-*self-similar*, if it is (r,α)-scaling for all $r > 0$.

Lemma 9.10. B^α *is α-self-similar and possesses stationary increments.*

Proof. $B^\alpha_{r,\alpha}$ and B^α_s are linear transformations of Gauss processes, i.e., Gauss processes, too. Therefore it suffices to show that the mean value and the covariance functions are the same as those of B^α. For $B^\alpha_{r,\alpha}$ we get $\mathbb{E}B^\alpha_{r,\alpha}(t) = r^{-\alpha}\mathbb{E}B^\alpha(rt) = 0$ and

$$\mathbb{E}(B^\alpha_{r,s}(s) - B^\alpha_{r,s}(t))^2 = \mathbb{E}(r^{-\alpha}B^\alpha(rs) - r^{-\alpha}B^\alpha(rt)))^2 = r^{-2\alpha}|rs - rt|^{2\alpha}$$
$$= |s - t|^{2\alpha},$$

which yields the equality for the covariance function. The arguments for B^α_s are similar and left to the reader. $\quad\square$

Remark 9.11. For $s, t \geq 0$ the random variable $B^\alpha(t+s) - B^\alpha(t)$ is normal distributed with density

$$p_s^\alpha(x) = \frac{1}{\sqrt{2\pi}s^\alpha} \exp\left(-\frac{x^2}{2s^{2\alpha}}\right).$$

The latter is uniformly bounded for $x \in \mathbb{R}$, $s \geq s_0$ and any fixed $s_0 > 0$.

The following general result will be used in order to determine the dimension of the graph of B^α. Here the potential theoretic method from Chapter 4 is applied.

Theorem 9.12. *Let X be a continuous stochastic process satisfying for certain $r, \alpha \in (0,1)$ the following:*
(1) X is (r, α)-scaling.
(2) The distributions of the random variables $X(t + s) - X(t)$ possess densities $p_{t,s}(x)$ with the property

$$\frac{1}{N}\int_0^N p_{t,s}(x)dt \leq \text{const}$$

for all $N > 0$, $x \in \mathbb{R}$, $s \in [cr, c)$ with some $c > 0$. Then

$$\dim_H \text{graph}(X) \geq 2 - \alpha \qquad \text{with probability } 1.$$

Proof. We first define the so-called *occupation measure* of X. (Recall that $[\Omega, \mathfrak{F}, \mathbb{P}]$ was the basic probability space.) For $\omega \in \Omega$ and Borel sets B in $\mathbb{R}^2$ set

$$\mu_\omega(B) := \int_0^T 1_B(t, X_\omega(t))dt.$$

Then μ is a random finite Borel measure concentrated on the graph of X. Below we will show that for any $1 < \beta < 2 - \alpha$ the mean β-energy of μ (cf. Definition 4.1) is finite, i.e.,

$$E^\beta := \mathbb{E} \int_{\mathbb{R}^2} \int_{\mathbb{R}^2} |x - y|^{-\beta} \mu(dx)\mu(dy) < \infty.$$

This implies

$$\int_{\mathbb{R}^2} \int_{\mathbb{R}^2} |x - y|^{-\beta} \mu_\omega(dx)\mu_\omega(dy) < \infty,$$

for all $\beta < 2 - \alpha$ and $\mathbb{P}$-almost all ω. From Corollary 4.7 we conclude for these ω,

$$\dim_H \text{graph}(X) \geq 2 - \alpha,$$

i.e., the assertion. It remains to show the finiteness of the mean β-energy. To this aim we estimate as follows.

$$\frac{1}{2}E^\beta = \frac{1}{2}\mathbb{E}\int_0^T\int_0^T \left((u-v)^2 + (X(u)-X(v))^2\right)^{-\beta/2} du\, dv$$

$$= \mathbb{E}\int_0^T\int_v^T \left((u-v)^2 + (X(u)-X(v))^2\right)^{-\beta/2} du\, dv$$

$$= \mathbb{E}\int_0^T\int_v^T \left(u^2 + (X(u+v)-X(v))^2\right)^{-\beta/2} du\, dv$$

$$\leq \int_0^T\int_0^T \mathbb{E}\left(u^2 + (X(u+v)-X(v))^2\right)^{-\beta/2} dv\, du$$

$$\leq \sum_{n=-k}^{\infty}\int_{r^{n+1}c}^{r^n c}\int_0^T \mathbb{E}\left(u^2 + (X(u+v)-X(v))^2\right)^{-\beta/2} dv\, du$$

(where k is the smallest natural number with $r^{-k}c > T$)

$$= \sum_{n=-k}^{\infty} r^n\int_{rc}^{c}\int_0^{Tr^{-n}} \mathbb{E}\left(r^{2n}s^2 + (X(r^n(s+t))-X(r^nt))^2\right)^{-\beta/2} dt\, ds$$

$$= \sum_{n=-k}^{\infty} r^n\int_{rc}^{c}\int_0^{Tr^{-n}} \mathbb{E}\left(r^{2n}s^2 + r^{2n\alpha}(X(s+t)-X(t))^2\right)^{-\beta/2} dt\, ds$$

(since X is (r,α)-scaling)

$$= \sum_{n=-k}^{\infty} r^n\int_{rc}^{c}\int_0^{Tr^{-n}}\int \left(r^{2n}s^2 + r^{2n\alpha}x^2\right)^{-\beta/2} p_{t,s}(x)dx\, dt\, ds$$

$$= T\sum_{n=-k}^{\infty} r^n\int_{rc}^{c}\int \left(r^{2n}s^2 + r^{2n\alpha}x^2\right)^{-\beta/2}\left[T^{-1}r^n\int_0^{Tr^{-n}} p_{t,s}(x)dt\right] dx\, ds$$

$$\leq \text{const} \sum_{n=-k}^{\infty} r^n\int_{rc}^{c}\int \left(r^{2n}s^2 + r^{2n\alpha}x^2\right)^{-\beta/2} dx\, ds$$

(since the expression in the brackets is bounded by a constant).
The last inner integral does not exceed

$$\int_{r^{2n}s^2 \geq r^{2n\alpha}x^2} \left(r^n s\right)^{-\beta} dx + \int_{r^{2n}s^2 < r^{2n\alpha}x^2} \left|r^{n\alpha}x\right|^{-\beta} dx$$

$$\leq \text{const}\, s^{1-\beta}r^{n(1-\alpha-\beta)}.$$

Therefore the above expression does not exceed

$$\text{const}\sum_{n=-k}^{\infty} r^n\, r^{n(1-\alpha-\beta)} < \infty,$$

since $r < 1$ and $2-\alpha-\beta > 0$. $\qquad\square$

The previous results enable us to determine the *dimensions of the graph of fractional Brownian motion.*

Corollary 9.13.

$$\dim_B \text{graph}(B^\alpha) = \dim_H \text{graph}(B^\alpha) = 2 - \alpha \ \ w.p.\,1\,.$$

Proof. By Lemma 9.7,

$$\dim_B \text{graph}(B^\alpha) \leq 2 - \alpha \ \ w.p.\,1\,.$$

Furthermore, Theorem 9.12 implies

$$\dim_H \text{graph}(B^\alpha) \geq 2 - \alpha \ \ w.p.\,1\,,$$

since the corresponding conditions are satisfied in view of Lemma 9.10 and Remark 9.11. Together with $\dim_H \leq \dim_B$ this yields the assertion. $\qquad\square$

Remark 9.14. The last result was first proved for the special case of the Wiener process in [Taylor (1955)] and then in [Adler (1977)] for more general Gaussian processes. In the last decades fractal properties of non-smooth stochastic processes and random fields have been studied intensively in the literature. An early example is [Xiao (1997)], where the gauge function for the exact Hausdorff measure of the graph of fractional Brownian motion was determined.

Bibliography

Adler, R. J. (1977). Hausdorff dimension and Gaussian fields, *Ann. of Probab.* **5**, 145–151.

Arbeiter, M. and Patzschke, N. (1996). Random self-similar multifractals, *Math. Nachr.* **181**, 5–42.

Bandt, C. and Graf, S. (1992). Self-Similar Sets 7. A characterization of self-similar fractals with positive Hausdorff measure, *Proc. Amer. Math. Soc.* **114**, 995–1001.

Barreira, L. (2011). *Thermodynamic Formalism and Applications to Dimension Theory*, Birkhäuser, Basel.

Bauer, H. (2001). *Measure and Integration Theory*, De Gruyter.

Beddord, T. and Fisher, A. M. (1992). Analogues of the Lebesgue density theorem for fractal sets and integers, *Proc. London Math. Soc.* **64**, 95–124.

Bishop, C. J. and Peres, Y. (1996). Packing dimension and Cartesian products, *Trans. Amer. Math. Soc.* **348**, 4433–4445.

Cohn, D. L. (2013). *Measure Theory*, Birkhäuser.

Cutler, C. D. (1995). The density theorem and Hausdorff inequality for packing measure in general metric spaces, *Illinois J. Math.* **39**, 676–694.

Dvoretzky, A. (1948). A note on Hausdorff dimension functions, *Proc. Camb. Phil. Soc.* **44**, 13–16.

Edgar, G. A. (1997). *Integral, Probability, and Fractal Measures*, Springer.

Elstrodt, J. (1996). *Maß- und Integrationstheorie*, Springer.

Falconer, K. (1985). Random fractals, *Math. Proc. Camb. Phil. Soc.* **100**, 559–582.

Falconer, K. (1990). *Fractal Geometry–Mathematical Foundations and Applications*, John Wiley.

Falconer, K. (1994). The multifractal spectrum of statistically self-similar measures, *J. Theor. Probab.* **7**, 681–702.

Falconer, K. (1997). *Techniques in Fractal Geometry*, John Wiley.

Federer, H. (1969). *Geometric Measure Theory*, Springer.

Frostman, O. (1935). Potentiel d'equilibre et capacité des ensembles avec quelques applications á la théorie des functions, *Meddel Lunds Univ. Mat. Sem.* **3**, 1–118.

Graf, S. (1987). Statistically self-similar fractals, *Probab. Th. Rel. Fields* **74**, 357–392.

Graf, S. (1995). On Bandt's tangential distribution for self-similar measures, *Monatsh. Math.* **120**, 223–246.

Haase, H. (1988), Packing measures on ultrametric spaces, *Studia Math.* **91**, 189–203.

Haase, H. (1990). The packing theorem and packing measure, *Math. Nachr.* **146**, 77–84.

Halmos, P. R. (1974) *Measure Theory*, Springer.

Hardy, G. H. (1916), Weierstrass's non-differentiable function, *Trans. Amer. Math. Soc.* **17**, 301–325.

Hausdorff, F. (1919), Dimension und äußeres Maß, *Math. Ann.* **79**, 157–179.

Howroyd, J. (1995). On dimension and on the existence of sets of finite positive Hausdorff measure, *Proc. London Math. Soc.* **70**, 581–604.

Hutchinson, J. E. (1981), Fractals and self similarity, *Indiana Univ. Math. J.* **30**, 713–747.

Joyce, H. (1999). A space on which diameter-type packing measure is not Borel regular, *Proc. Amer. Math. Soc.* **127**, 985–991.

Joyce, H. and Preiss, D. (1995). On the existence of subsets of finite positive packing measure, *Mathematika* **42**, 14–24.

Kaplan, H., Mallet-Peret, J. and Yorke, J. A. (1984). The Lyapunov dimension of a nowhere differentiable attracting torus, *Ergod. Th. and Dynam. Syst.* **4**, 261–281.

Klenke, A. (2020). *Probability Theory — A Comprehensive Course*, 3rd ed., Springer.

Mandelbrot, B. (1982). *The Fractal Geometry of Nature*, W.H. Freeman and Co.

Marstrand, J. M. (1964). The (φ, s)-regular subsets of n-space, *Trans. Amer. Math. Soc.* **113**, 369–392.

Mattila, P. (1995). *Geometry of Sets and Measures in Euclidean Spaces*, Cambridge University Press.

Mattila, P. and Mauldin, D. (1997). Measure and dimension functions: measurability and densities, *Math. Proc. Camb. Phil. Soc.* **121**, 81–100.

Mauldin, R. D. and Urbanski, M. (2003). *Graph Directed Markov Systems: Geometry and Dynamics*, Cambridge University Press.

Mauldin, R. D. and Williams, S.(1986). Random recursive constructions, *Trans. Amer. Math. Soc.* **295**, 325–346.

May, R. M. (1976). Simple mathematical problems with very complicated dynamics, *Nature* **261**, 459–467.

Moran, P. A. P. (1946). Additive functions of intervals and Hausdorff measure, *Math. Proc. Camb. Phil. Soc.* **42**, 15–23.

Olsen, L. (2000). Multifractal geometry, In: *Fractal Geometry and Stochastics, II. Eds.: C. Bandt, S. Graf and M. Zähle. pp.3–37*, Progr. Probab. 46, Birkhäuser, Basel.

Patzschke, N. and Zähle, U. (1990). Self-similar random measures. IV. The recursive construction model of Falconer, Graf, and Mauldin and Williams, *Math. Nachr.* **149**, 285–302.

Patzschke, N. and Zähle, M. (1994). Fractional differentiation in the self-affine case. IV–Random measures, *Stochastics and Stochasics Reports* **49**, 87–98.

Pesin, J. (1997). *Dimension Theory in Dynamical Systems*, The University of Chicago Press.

Pesin, J. and Climenhaga, V. (2009). *Lectures on Fractal Geometry and Dynamical Systems*, Amer. Math. Soc.

Pontryagin, L. S. and Shnirelman, L. G. (1932). Sur une Propriete Metrique de la Dimension, *Ann. Math.* **33**, 156-162.

Preiss, D. (1987). Geometry of measures in $\mathbb{R}^n$: distribution, rectifiablity, and densities, *Ann. of Math.* **125**, 537–643.

Przytycki, F. and Urbanski, M. (2010). *Conformal Fractals: Ergodic Theory Methods*, Cambridge Univ. Press.

Rogers, C. A. (1998). *Hausdorff Measures*, Cambridge University Press.

Saint Raymond, X. and Tricot, C. (1988). Packing regularity of s-sets in n-space, *Math. Proc. Camb. Phil. Soc.* **103**, 133–145.

Schief, A. (1994). Seperation properties for self-similar sets, *Proc. Amer. Math. Soc.* **122**, 111–115.

Schief, A. (1996). Self-similar sets in complete metric spaces, *Proc. Amer. Math. Soc.* **124**, 481–490.

Shen, W. (2018). Hausdorff dimension of the graphs of the classical Weierstrass functions, *Mathematische Zeitschrift* **289**, 223–266.

Taylor, S. J. (1955). The α-dimensional measure of the graph and set of zeros of a Brownian path. *Math. Proc. Cambridge Philos. Soc.* **51**, Part II, 265–274.

Taylor, S. J. and Tricot, C. (1985). Packing measure and its evaluation for a Brownian path, *Trans. Amer. Math. Soc.* **288**, 679–699.

Tricot, C. (1982). Two definitions of fractal dimension, *Math. Proc. Camb. Phil. Soc.* **91**, 57–74.

Tudor, C. A. (2014). *Analysis of Variations for Self-similar Processes*, Probability and its Applications, Springer, Cham.

Walters, P. (1982). *An Introduction to Ergodic Theory*, Springer.

Xiao, Y. (1997). Hausdorff measure of the graph of fractional Brownian motion, *Math. Proc. Camb. Phil. Soc.* **122**, 565–576.

Zähle, U. (1988). Self-similar random measures. I. Notion, carrying Hausdorff dimension, and hyperbolic distribution, *Probab. Theory Related Fields* **80**, 79–100.

Zähle, M. (2001). The average density of self-conformal measures, *J. London Math. Soc.* **63**, 721–734.

Index

Nomenclature

Lectures on Fractal Geometry

$\mathcal{S}^s$	s-dimensional spherical measure, page 15
$\mathfrak{X}_\mu$	family of μ-measurable subsets of the space X, page 3
$\mu \llcorner A$	restriction of the measure μ to the set A, page 3
μ^τ	metric outer covering measure with weight function τ, page 11
$\overline{\dim}\mu(x)$	upper pointwise dimension of the measure μ, page 69
$\overline{C}$	closure of the set C, page 14
$\underline{\dim}\mu(x)$	lower pointwise dimension of the measure μ, page 69
d_H	Hausdorff distance between compact sets, page 76
n_δ	covering number, page 57

Printed in the United States
by Baker & Taylor Publisher Services